MARCH 2017

Defense Acquisition Trends, 2016

The End of the Contracting Drawdown

Project Director
Andrew P. Hunter

Project Codirector
Rhys McCormick

Principal Authors
Jesse Ellman
Samantha Cohen
Andrew Hunter
Kaitlyn Johnson
Rhys McCormick
Gregory Sanders

Contributing Authors
Gabriel Coll
Loren Lipsey

A Report of the CSIS Defense360 Series on Strategy, Budget, Forces, and Acquisition

CSIS | CENTER FOR STRATEGIC & INTERNATIONAL STUDIES

ROWMAN & LITTLEFIELD

Lanham • Boulder • New York • London

About CSIS

For over 50 years, the Center for Strategic and International Studies (CSIS) has worked to develop solutions to the world's greatest policy challenges. Today, CSIS scholars are providing strategic insights and bipartisan policy solutions to help decision-makers chart a course toward a better world.

CSIS is a nonprofit organization headquartered in Washington, D.C. The Center's 220 full-time staff and large network of affiliated scholars conduct research and analysis and develop policy initiatives that look into the future and anticipate change.

Founded at the height of the Cold War by David M. Abshire and Admiral Arleigh Burke, CSIS was dedicated to finding ways to sustain American prominence and prosperity as a force for good in the world. Since 1962, CSIS has become one of the world's preeminent international institutions focused on defense and security; regional stability; and transnational challenges ranging from energy and climate to global health and economic integration.

Thomas J. Pritzker was named chairman of the CSIS Board of Trustees in November 2015. Former U.S. deputy secretary of defense John J. Hamre has served as the Center's president and chief executive officer since 2000.

CSIS does not take specific policy positions; accordingly, all views expressed herein should be understood to be solely those of the author(s).

Acknowledgments

This report was made possible by general support to the CSIS Defense-Industrial Initiatives Group and International Security Program.

ISBN: 978-1-4422-8011-3 (pb); 978-1-4422-8012-0 (eBook)

Center for Strategic & International Studies
1616 Rhode Island Avenue, NW
Washington, DC 20036
202-887-0200 | www.csis.org

Rowman & Littlefield
4501 Forbes Boulevard
Lanham, MD 20706
301-459-3366 | www.rowman.com

Contents

Figures

Tables

Executive Summary

This report, *Defense Acquisition Trends, 2016: The End of the Contracting Drawdown*, is the second in an annual series of reports examining trends in what DoD is buying, how DoD is buying it, and whom DoD is buying from. The *Defense Acquisition Trends* reports are part of a broader series of reports titled "Defense Outlook: A CSIS Series on Strategy, Budget, Forces, and Acquisition." This year's report looks in great depth at issues in research and development, acquisition reform in the FY2017 National Defense Authorization Act (NDAA), Performance of the Defense Acquisition System, the future of cooperative International Joint Development Programs, and major trends apparent in the activities of the major defense components. By combining detailed policy and data analysis, this report provides a comprehensive overview of the current and future outlook for defense acquisition.

Since the peak in 2009, the main stories in defense acquisition have been variations on a theme of downturn: the postwar budget drawdown, defense budget caps, and sequestration and its aftermath. The last several years have seen a consistent decline in DoD contract obligations, beyond the overall decline in total DoD obligations, leading to a significant reduction in contract obligations as a share of total net obligations as seen in Figure I:

Figure I: Defense Contract Obligations vs. Total Defense Net Obligations, 2008–2015[1]

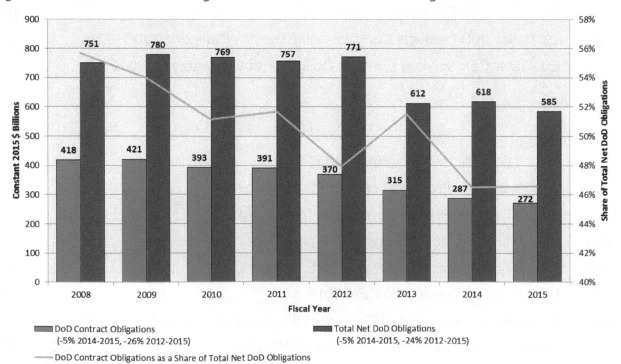

Source: Federal Procurement Data System (FPDS); DoD Comptroller Financial Summary Tables; CSIS analysis.

---

[1] See Appendix A, Methodology, for a detailed description of how CSIS calculated total DoD net obligations.

While DoD contract obligations have declined steadily since 2009, with an acceleration between 2012 and 2013 due to the impact of sequestration, total DoD net obligations were largely steady until 2013. While contract obligations have declined more steeply than total net obligations since 2009, since 2012, the rates of decline for contract obligations and total net obligations have been roughly equivalent. This is particularly apparent in 2015, where both contract obligations and total net obligations declined by 5 percent. This leveling-off appears to indicate that the period of disproportionate declines in DoD contract obligations may be at the end. And while defense budgets and contract obligations are now beginning to increase again, yielding an end to the contracting drawdown, it is not clear whether DoD contract obligations will regain their lost ground as a share of total net obligations in the recovery that follows.

With FY2016 data available and complete in the Federal Procurement Data System (FPDS) as of January 2017, however, CSIS was able to establish that the tide has definitely turned in the direction of contract spending. Our initial analysis of the FY2016 data shows that overall DoD contract obligations rose by 7 percent in 2016, far higher than predicted. Figure II shows the dimensions of this increase, broken down by major DoD component:

Figure II: Defense Contract Obligations by Component, 2009–2016

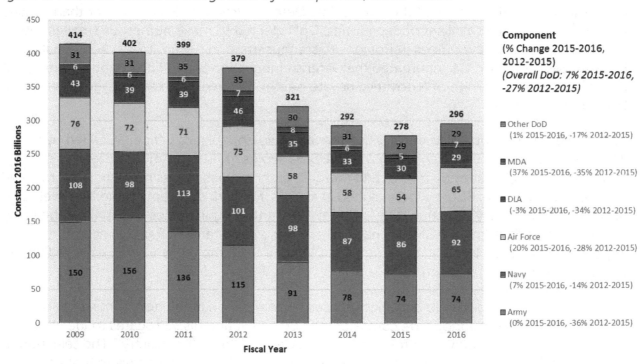

Source: FPDS; CSIS analysis.

Missile Defense Agency (MDA), Air Force, and the Navy all saw significant increases in contract obligations in 2016, driven primarily by increased obligations for large procurement programs like the C-130J transport aircraft, the KC-45A tanker aircraft, and the Trident II missile program. Even the Army, which had declined far more steeply than DoD overall throughout the budget drawdown, was virtually stable between 2015 and 2016.

While CSIS continues to analyze the just-available FY2016 DoD contract data, the bulk of this report focuses on a deep analysis of contract trends in the years prior to 2016, as well as policy trends through 2016. CSIS's findings on the key issues in defense acquisition are organized in four main sections:

- What is DoD buying?

- How is DoD buying it?

- Whom is DoD buying from?

- What are the defense components buying?

What Is DoD Buying?

Defense Innovation Initiative

Over the past year, activities associated with the Defense Innovation Initiative remained a top DoD priority. In the PB17 budget, DoD included $18 billion for Third Offset Strategy investments over the course of the Future Year Defense Program (FYDP). Rather than making large initial bets on a small set of capabilities, DoD elected to make numerous smaller bets. DIU(X) opened two new offices in Boston, Massachusetts, and Austin, Texas. Additionally, Defense Secretary Ash Carter created the Defense Innovation Board composed of a range of experts to identify a range of innovative private-sector practices and technological solutions that DoD can adapt.

In 2016 Secretary Carter also declassified the Strategic Capabilities Office (SCO) led by Dr. Will Roper. Originally established in 2012, the SCO works to ensure America's lead in military technological capabilities by mixing and merging technologies across multiple platforms and services. For example, SCO's work resulted in successful work on adapting the hypervelocity projectile for the Navy's existing conventional naval artillery gun pieces and making the Army Tactical Missile System (ATACMS) capable of targeting naval targets. Perhaps most importantly, the SCO has achieved significant "buy in" from the military services as reflected by its increasing budgets each year.

At this time, the prospects for activities focused on innovation in the new administration remains unclear. Throughout the campaign, the president and his surrogates made little reference to military innovation focusing instead on "rebuilding the military." The selection of retired Gen James Mattis as secretary of defense provides little guidance given that he focused mostly on operational activities throughout his distinguished military career. However, the decision to retain, at least temporarily, Deputy Secretary Bob Work suggests that military innovation might remain a DoD priority. Whomever is selected to eventually replace Deputy Secretary Work will be an indicator about the long-term future of activities associated with the Defense Innovation Initiative.

Procurement Contracts Drive Slowing of Decline in 2015, Increase in 2016

As discussed above, the decline in DoD contract obligations through 2015 was not evenly distributed across the range of what DoD contracts for, and this uneven distribution remains the case in the initial year of the contracting recovery, as seen in Figure III:

Figure III: Overall Defense Contract Obligations by Area, 2000–2016

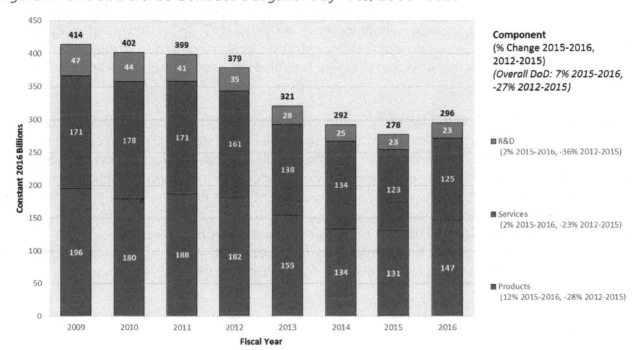

Source: FPDS; CSIS analysis.

The leveling-off of the decline in overall DoD contract obligations is primarily driven by DoD products contracts, which were stable (0 percent) in 2015, after declining sharply in 2014 (-14 percent) and in parallel to overall DoD contracts since 2012. Year-to-year trends in products contract obligations are highly sensitive to the timing of large contracts for production of major weapons systems such as the F-35, though with purchases of F-35s likely to accelerate in the near future, these large contracts may be a continuing source of stability within the DoD products contract portfolio. The increase in DoD contract obligations in 2016 was largely driven by increases in obligations for procurement of major weapons systems, though obligations for the F-35 actually declined in 2016.

Trough in Development Pipeline for Major Weapons Systems Continues into Seventh Year

As many analysts and policymakers feared, DoD R&D contracts have borne a disproportionate share of cuts within the DoD contracting portfolio during the current budget drawdown. The dimensions of those cuts, however, have not followed the expected path. Despite fears that early stage, seed corn R&D would be hit particularly hard, the data show that it has been relatively preserved compared to the overall declines in R&D. In fact, within DoD, two categories of mid- to late-stage R&D, Advanced Technology Development (6.3) and System Development & Demonstration (6.5), have seen cuts of two-thirds or more between 2009 and 2015.

The two main drivers of the massive declines in those two stages of R&D are the cancelation of large R&D programs (such as the Army's Future Combat Systems) and the maturation of R&D programs into procurement (such as the F-35 Joint Strike Fighter). During the budget drawdown period, however, there has been a dearth of new development programs for major weapons systems that replace those that have either graduated into production or been canceled. In 2016, despite a slight increase in overall DoD R&D contract obligations, obligations for System Development & Demonstration (6.5) continued to decline sharply. As a result, DoD is facing what is now a seven-year trough in its development pipeline for major weapons systems.

This trough has manifested differently within the three military services. In the Air Force, significant work and funding for the B-21 bomber is likely to begin in the next couple of years. The Navy has the Columbia-class ballistic missile submarine program on the horizon, and with the program recently receiving Milestone B certification, contract obligations for development of the submarine should reach significant levels in the next few years. The Army is in the toughest position of the three, as since the failure of Future Combat Systems, the Army has been largely unable to start or sustain major development programs. With continuing uncertainty about future missions and capabilities, as well as significant budgetary challenges, the Army's trough seems likely to persist for the foreseeable future.

How Is DoD Buying It?

Acquisition Reform in the FY2017 NDAA

Between the House and Senate versions of the FY2017 NDAA, Congress enacted fundamental changes across three major elements of the defense acquisition system: how the system is organized and given its mission; how acquisition programs are structured; and what the business model is for defense research and development. The final conference agreement largely adopted House and Senate proposals to shift away from the traditional Major Defense Acquisition Program (MDAP) structure around which much of the defense acquisition system has centered, creating new openings for more prototyping, technology demonstrations, and acquisitions of commercial technology. This shift aligns with the decision to divide the responsibilities of the current USD AT&L between a USD R&E focused on early-stage technologies, and potentially on new business models for R&D, and a USD A&S focused on traditional MDAP programs. However, many of the statutory changes enacted will not be fully implemented until 2018, and the desire to reshape the defense acquisition system in Congress combined with the Trump administration's early interest in the topic suggests the likelihood for continued debate on these fundamental issues in the coming years. The new administration will need to quickly grapple with the changes in the FY2017 NDAA and determine how to balance the more incremental, internal DoD approach to acquisition improvement embodied in Better Buying Power with the more fundamental shifts desired by Congress, and incorporate these approaches into its own agenda for defense acquisition.

Performance of the Defense Acquisition System

The performance of the defense acquisition has been hotly debated at many points in U.S. history. Recent years are no exception, as exemplified by the Government Accountability Office (GAO) listing the system for major weapons buying as "high risk" for nearly a quarter of a century.[2] Thanks to longstanding transparency efforts, many of the basic facts are not in contention, but different reformers have different priorities and naturally favor different metrics. For example, Sen. John McCain points to the enormous total cost growth since the start of each program in the portfolio, which can mean going back to the 1990s. By comparison, Under Secretary Kendall favors year-on-year metrics, which show improvement.

Recent studies by other researchers, presented at CSIS and the Naval Postgraduate School's Defense Acquisition Research forum, have particularly illuminated two metrics: cost and schedule growth. This research focuses on performance of major defense acquisition programs where such data is routinely gathered. The good news is that Under Secretary Kendall's analysis shows similar trends as the GAO report, namely that DoD's Better Buying Power (BBP) initiatives appear to be bearing fruit and the ratio of actual unit costs relative to original estimates is improving with "a net buying power gain of $10.7 billion" in 2015.[3] This progress is particularly impressive, given research from the Institute for Defense Analyses (IDA) that finds that, with the exception of the Packard commission reforms of 1969, better cost performance is associated with periods of growing budgets rather than being associated with periods of major acquisition reform. Based on this IDA finding and the major budget drawdown of the last six years, it is remarkable that acquisition system cost performance has actually improved. The bad news is that the schedule growth is getting worse, although IDA research indicates that this is attributable to unrealistic estimates rather than increases in project cycle time.[4]

CSIS extended the scope of research on the performance of the defense acquisition system by looking at defense contracts as a whole using the metrics of partial or complete contract terminations as well as cost-ceiling breaches. The preliminary outcomes on contract terminations indicates that shorter-duration contracts begun as budgets began to decline experienced a bump in terminations but that longer-duration contracts begun in recent years have been less likely to be terminated. The results on ceiling breaches were more straightforward: the period after the BBP reforms were initiated coincided with a noticeable decline in ceiling breaches. Taken together, this research suggests that real progress is being made on cost performance, and that to the extent this remains a priority, future reform efforts would benefit from building on these successes. However, those seeking on-time delivery or a more agile acquisition system have a heavier lift and should be sure to address the system's current weaknesses in schedule estimation in their plans.

---

[2] Michael J. Sullivan, *Weapon Acquisition Program Outcomes and Efforts to Reform DOD's Acquisition Process*, May 1, 2016, 5, http://www.acquisitionresearch.net/publications/detail/1698/.

[3] Ibid., 12.

[4] David M. Tate, *Acquisition Cycle Time: Defining the Problem*, Institute for Defense Analyses, October 2016, 78–79, https://www.ida.org/idamedia/Corporate/Files/Publications/IDA_Documents/CARD/2016/D-5762.pdf.

Fixed Price/Cost Reimbursement Balance Largely Stable, But Major Shifts in Fee-Type Usage

During the budget drawdown, the usage levels of fixed price and cost reimbursement contract types in DoD contracting have been largely unchanged. This stability is broad-based across most of the major DoD components and across the range of what DoD contracts for, with usage within the products, services, and R&D contracting portfolios largely unchanged during the budget drawdown period.

There have been significant shifts, however, in the fee types used for both fixed price and cost reimbursement contracts within DoD. The importance of fee type has been shown in Under Secretary Kendall's 2014 Performance of the Defense Acquisition Report, which found no performance difference between fixed price and cost reimbursement, but significant benefits from use of incentive-fee contract types. Putting this finding into action, the share of fixed price contract obligations structured as Fixed Price Incentive Fee has risen from 2 percent in 2008 to 14 percent in 2014 and 2015, primarily within the DoD products and R&D contracting portfolios. Surprisingly, there has been no corresponding increase in the use of Cost Plus Incentive Fee contract types; the share of cost reimbursement contract obligations structured as Incentive Fee has fallen from 23 percent in 2011 to 14 percent in 2015, primarily within products contracts. Additionally, the share of cost reimbursement contract obligations structured as Cost Plus Award Fee, which hovered between 40 percent and 49 percent from 2000–2007, has declined steadily since, to 11 percent in 2015. In its place, use of Cost Plus Fixed Fee has risen from 36 percent of cost reimbursement contract obligations in 2007 to 67 percent in 2015.

Looking at contract type usage for contracts funded out of the different DoD budget accounts, a few notable trends and data points emerge. First, for products and services contracts funded out of the RDT&E account, which collectively account for nearly half of contracts funded out of that account, far greater shares are structured as cost reimbursement than for products and services in DoD overall. Second, the rising usage of Cost Plus Fixed Fee is particularly stark within O&M; nearly three-quarters of cost-reimbursement services contract obligations funded out of O&M are now structured as Cost Plus Fixed Fee, up from less than half as recently as 2012. And third, for services contract obligations funded out of Procurement, the share structured as cost reimbursement is 19 percentage points higher than for overall DoD services contracts.

From Whom Is DoD Buying?

Small Vendors Gain Share during Budget Drawdown, Big 5 Share of DoD R&D Continues to Plummet

Despite the massive decline in DoD contract obligations since 2009, the composition of the defense industrial base, as measured by size of vendor, has been relatively stable in the 2009–2015 period. Medium vendors have accounted for between 20 percent and 22 percent of overall DoD contract obligations in every year during the period, while Large vendors have accounted for between 30 percent and 34 percent throughout. The Big 5 defense vendors[5] have seen similar stability, accounting for between 27 percent and 31 percent of DoD contract obligations in each year since 2009.

Small vendors, by contrast, have actually increased their share in the last two years, from 16 percent in 2009–2013 to 19 percent in 2014 and 2015; this increase is broad-based, across most of the major DoD components and within both the services and R&D contracting portfolios. The bulk of this growth in share was a result of an increase in small business contract obligations between 2013 and 2014, after the tremendous decline in 2013 in the wake of sequestration. Obligations to Small vendors declined slightly in 2015, but nonetheless, Small is the only category of vendors for which 2015 contract obligations are higher than they were in 2013. In 2016, obligations to Small vendors increased roughly in line with the overall increase in DoD contract obligations, and Small vendors maintained their share of the overall DoD contracting market. This data can be seen as a victory for policies that promote Small business participation: despite the pressures of the budget drawdown, Small vendors have managed to not just maintain their place in the defense contracting marketplace, but increase it.

Of the three areas of the defense-contracting marketplace, R&D has seen by far the most dramatic shift in the composition of the supporting industrial base. In 2009, the Big 5 vendors accounted for 57 percent of DoD R&D contract obligations, but that share has declined steadily since, to just 33 percent in 2015, and further declined to 29 percent in 2016. This decline is particularly acute within the Army's R&D contracting portfolio, where the share of contract obligations going to the Big 5 having fallen from 48 percent in 2009 to 18 percent in 2015, and dropping further in 2016, to just 5 percent. This overall decline is particularly notable because of the massive decline in DoD R&D contract obligations since 2009; overall, the Big 5 control roughly two-fifths less of a market that is less than half the size it was in 2009. The primary driver of this decline is the now-seven-year trough in DoD's development pipeline for major weapons systems that was discussed briefly in the "What Is DoD Buying?" section of this Executive Summary. With a number of major development programs either maturing into production or getting canceled, and a dearth of new large development programs starting up, the high-value defense R&D contracting marketplace has shrunk significantly across the major DoD components.

Relative Stability in Services Industrial Base despite Wave of M&A Activity

Looking at DoD services vendors, there have been only minor shifts in the composition in the industrial base, with the most notable being an increase in the share of services contract obligations going to Small vendors, from 21 percent in 2009 to 26 percent in 2015, with significant increases in three of the five categories of services. This relative stability over the period, particularly over the last two years, may be somewhat surprising given the wave of M&A activity within the government services sector in recent years. The overall trend with the government services market is twofold: first, a trend of diversified vendors divesting their government services business units (particularly in government IT services); and second, government services-focused vendors merging with or acquiring other vendors to increase

---

[5] Lockheed Martin, Boeing, Northrop Grumman, Raytheon, and General Dynamics.

market share and access to markets/sectors. These changes, however, have not yet been reflected in the data on the composition of the DoD services industrial base.

Concentration of Overall Defense Industrial Base Largely Stable during Drawdown, But Significant Shifts within Products and R&D

To study trends in the concentration of the defense industrial base across the current budget drawdown, CSIS looked at the top 20 prime vendors (measured by prime contract obligations) in 2009 and 2015), and calculated the shares of contract obligations going to the top 5 and top 20 vendors. For DoD overall, the data show no significant shift in the concentration of the defense industrial base overall since 2009: the share of total DoD contract obligations going to the top 5 vendors was virtually the same in 2009 (27 percent) as in 2015 (28 percent), and the same is true when looking at the share of overall DoD contracts going to the top 20 vendors in 2009 (44 percent) and 2015 (45 percent.)

There has been a significant increase, however, in the concentration of the DoD products market among the top 5, with the share of overall DoD products contract obligations going to the top 5 rising from 34 percent in 2009 to 42 percent in 2015. Similarly, the share captured by the top 20 vendors rose from 57 percent in 2009 to 63 percent in 2015.

The rising concentration in products was offset by the massive decline in the concentration of the DoD R&D industrial base, due to the seven-year trough discussed in the "What Is DoD Buying?" section above. The top 5 share of overall DoD R&D contract obligations has declined from 57 percent to 37 percent, and the top 20 share has fallen from 77 percent to 65 percent. Since the largest vendors disproportionately perform the largest R&D projects, it is not surprising that a dearth of large development programs would drastically reduce the share of R&D going to those large prime vendors.

What Are the Defense Components Buying?

Massive Decline in Army Contracts Begins to See Bottom in 2015

Of the three military services, the Army has been hit hardest by the declines in DoD contract obligations since 2012, falling by 36 percent over the 2012–2015 period, notably more steeply than overall DoD contract obligations. 2014 saw a 14 percent decline within the Army's contracting portfolio, which was also steeper than the decline for overall DoD, but in 2015, Army contract obligations declined by only 6 percent, which was roughly in line with the overall decline in DoD contract obligations. This deceleration of the decline in Army contract obligations was driven by relative stability within the Army's products and R&D contracting portfolios between 2014 and 2015; by contrast, Army services contracts (-11 percent) declined at nearly twice the rate of overall Army contracts in 2015, and more steeply than services contracts in DoD overall.

The acceleration of the decline in services contracts in 2015 may presage further declines within the Army's services contracting portfolio in the near term. But the overall slowing of the decline indicates that the steep decline in Army contract obligations since 2009, driven by the winding-down of combat operations in Iraq and Afghanistan and the Army's recent inability to start and sustain major development programs, may be close to reaching its bottom. The overall stability in Army contract obligations in 2016 lends further evidence to support this hypothesis.

Rate of Competition for Air Force Services Contracts Continues to Plummet

In an October 2015 report on defense competition, CSIS noted that the Air Force has seen a significant decline in its level of effective competition[6] for services contract obligations, in a period where competition rates for services in other parts of DoD are either stable or rising. For overall Air Force services, the rate of effective competition (54 percent) was already notably lower than the rate for non-Air Force services (70 percent). That gap has only widened in recent years: while the rate of effective competition for non-Air Force services has risen slightly, to 72 percent in 2015, the rate for Air Force services has fallen to 39 percent in 2015.

Even when looking at the specific types of services (such as Maintenance/Repair of Aircraft or Engineering & Technical Services) that the Air Force contracts for, the Air Force consistently sees lower (and often declining) rates of effective competition for the same types of services than does the rest of DoD. This trend continues to surprise the study team, because the Air Force has been seen as taking a leading role in improving tradecraft in the acquisition of services.

Final Thoughts

The end of 2016 is ushering in a set of major shifts in defense acquisition. The end of the contracting drawdown, the adoption of fundamental shifts in acquisition as a result of congressional action, and the beginning of a new administration with new priorities all herald the likelihood of major changes in 2017. In this report, CSIS seeks to provide an empirical picture of the system as it stands now that incorporates the accumulated result of top-down decision-making and bottom-up implementation. This picture demonstrates that policy initiatives undertaken with respect to defense acquisition do not always conform to expectations. The fact that R&D contracting has dropped dramatically, in the midst of a seven-year trough in new program starts, but is not eating its seed corn, shows that how policies are implemented can dramatically shape the defense acquisition system.

In other cases policy and implementation show a clear disconnect. Competition rates for Air Force services continue to decline despite Air Force leadership in service acquisition tradecraft. Likewise, despite a finding that incentive fee contracts have better cost performance, the usage rate for cost plus incentive fee contracts have declined with cost plus fixed fee on the rise instead.

Finally, research by GAO and CSIS indicates one area where top level policy appears to be showing sustained success. Namely the Better Buying Power initiative appears to be bearing fruit with lower cost overruns for both major weapon systems and defense contracts overall.

---

[6] CSIS defines a contract as effectively competed if it is competitively sourced and receives at least two offers. See Appendix A, Methodology, for more details on how CSIS analyzes trends in competition for DoD contracts.

With scrutiny of the defense acquisition system unlikely to disappear in coming years, the ability to accurately estimate and then deliver to cost targets will remain a critical measure of success. However, other measures of success, such as responding quickly to a changing security environment, are also priorities for the defense acquisition system. Improving performance on these measures of success without sacrificing hard won gains in cost performance will require dedicated, informed leadership.

1. Introduction

This report represents the second edition of our annual series of reports examining trends in defense acquisition, which follow on and expand the work of CSIS's earlier annual series of reports on defense contract trends. It is part of a larger series of reports issued by CSIS that are intended to provide an annual look at developments in strategy, budget, forces, and acquisition that provide the reader with a comprehensive and up-to-date understanding of major issues impacting the nation's defense.[7]

The defense acquisition system experienced a unique moment in 2016. As this year's report confirms, the long drawdown in the defense acquisition system that began in 2009 hit bottom in 2015, and as of 2016 the size of the defense acquisition system is now growing again. At the same time, there are numerous predictions from experienced observers that the defense acquisition system is in the midst of a once-in-a-generation shift in the nature of what it is buying, the acquisition mechanisms needed to fulfill DoD's needs efficiently, and the shape of the industrial base supporting the Department of Defense.[8] The extent to which this shift requires changes in the way that DoD organizes for and executes acquisition programs, with major changes being pursued by Congress and in some areas promoted and in other areas resisted by the Department of Defense, is explored fully in this report.

In addition, the close of 2016 brought the advent of a new administration that has shown every indication of being an early and active participant in the debate over defense acquisition. By combining much of CSIS's ongoing analysis of current defense acquisition policy debates with detailed data on what is really going on in the acquisition system in one report, the study team hopes to make both the public and policymakers better able to follow and participate in these debates.

1.1. Report Organization

At the beginning of a new administration that has committed to providing increased support to the military and also to applying close scrutiny to the performance of the defense acquisition system, this report assesses the state of defense acquisition system it is inheriting. To begin our analysis, the initial two chapters look at the current context of the defense acquisition system:

- **Chapter 2: An Initial Look at FY2016 Contract Trends** provides an overview of the most recent contract data, focusing on a few key developments relating to the issues and trends developed in CSIS's extensive analysis of FY2015 and previous year contract data.

---

[7] The complete series of these reports can be found on CSIS's Defense360 site under the heading "Defense Outlook," https://defense360.csis.org/.

[8] See, for example, the discussion at CSIS's annual Global Security Forum 2016 panel on "Defense Market Outlook: Challenges for the Next Administration," https://www.csis.org/events/global-security-forum-2016-defense-market-outlook-challenges-next-administration.

- **Chapter 3: DoD Contract Spending in a Budgetary Context** establishes the larger budgetary context in which the acquisition trends identified in this report are occurring.

Subsequent chapters dive in detail into a major single research question, in most cases using data available in 2016 (from FY2015 and earlier years) and a handful of related questions:

- **Chapter 4:** What Is DoD Buying?
  How have the defense drawdown and budget caps changed what DoD is buying? How is DoD's focus on innovation faring? Given the recent focus on the erosion of U.S. technological superiority in key mission areas, how is DoD's investment in research and development keeping pace?

- **Chapter 5: How Is DoD Buying It?**
  What major acquisition reform efforts are currently underway? How have DoD contracting approaches changed over time? What performance metrics can be derived from publicly available DoD contract data? What changes are occurring in DoD's approach to contracting for services?

- **Chapter 6: Whom Is DoD Buying From?**
  How has the composition of prime vendors changed during the drawdown and what causes can be identified? Who are the top vendors and what do they tell us about industrial base consolidation? What's the baseline for DoD outreach for Silicon Valley?

- **Chapter 7: What Are the Defense Components Buying?**

  How have the budget drawdown, sequestration, and its aftermath affected contract spending within the major DoD components? What are the specific sources of any increases or declines in contract obligations within the major DoD components?

- **Chapter 8: Conclusion**

  The final chapter of the report summarizes the major findings of the study team to these questions.

The report's research approach is discussed in Appendix A: Methodology.

2. An Initial Look at FY2016 Defense Contract Trends

Though this report focuses on a deep dive into the data on DoD contracting through FY2015, as of the beginning of January 2017, full FY2016 contract data for DoD is now available through FPDS. While the data will likely continue to be updated in the coming months, and the total contract dollars for FY2016 are likely to increase as straggler data makes its way into the database, the available data is complete and reliable enough to provide an initial look into the most recent available data on the state of DoD contracting.

2.1. The Drawdown of DoD Contracts Ends in Fiscal Year 2016

Based on the continued slowing of the decline in DoD contract obligations observed in 2015, both for DoD overall and for some of its major components, CSIS had predicted the decline was close to reaching its floor. In fact, the picture is far better than most anyone would have predicted, as overall DoD contract obligations increased by 7 percent, representing an increase of $18 billion over 2015 obligations levels. Figure 2-1 shows this increase, broken down by major DoD component:

Figure 2-1: Defense Contract Obligations by Component, 2009–2016

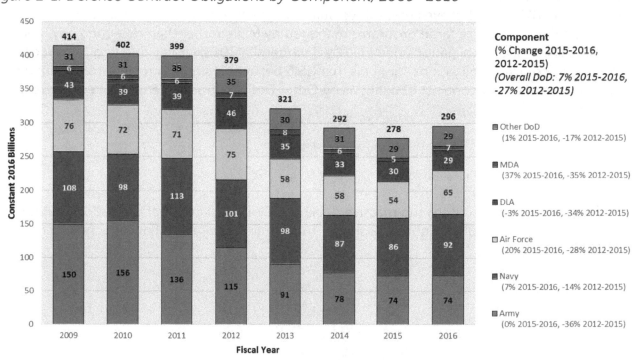

Source: FPDS; CSIS analysis.

The most striking feature of the 2016 DoD contract obligations data is the leveling off of the decline in Army contract obligations. The Army has borne the brunt of the budget drawdown, due to the withdrawal from Iraq and winding down of operations in Afghanistan, as well as the lack of new major procurement programs within the Army. Just between 2012 and 2015, Army contract obligations declined by 36 percent, notably faster than the overall decline in DoD contract obligations. But after declining by only 5 percent in 2015, Army contract obligations were virtually stable in 2016. While there were significant changes within specific parts of the Army's contracting portfolio (~$2 billion increase in obligations for Aircraft, ~$1 billion decline in obligations for Missiles & Space), the overall picture indicates that, while a significant increase in Army contract obligations is unlikely for the foreseeable future, the period of rapid decline is likely at an end.

The Air Force, which saw declines between 2012–2015 that were roughly in line with the overall decline in DoD contract obligations, saw a massive 20 percent increase in 2016, representing growth of $11 billion over 2015 obligations levels. The key drivers of this increase were major increases in obligations for two Aircraft programs: the C-130J transport aircraft ($3.8 billion increase) and the KC-45A tanker aircraft ($2.5 billion increase). The Navy, which saw its contract obligations decline at roughly half the rate of overall DoD between 2012 and 2015, saw a modest 7 percent increase in contract obligations in 2016. This increase was driven by increasing obligations for the P-8A Poseidon ($1.7 billion increase) and E-2C Hawkeye (~$500 million) aircraft programs, the LHA amphibious assault ship (~$950 million), and nuclear components for Trident II missiles (~$800 million), offsetting declines in obligations for the F-35 Joint Strike Fighter[9] ($2.2 billion decline) and Trident II missiles ($1 billion decline).

MDA contract obligations increased by 37 percent in 2016, but this follows a 24 percent decline in 2015; CSIS believes that this volatility is likely the result of the timing of contracts for the small number of large programs managed by MDA, rather than reflecting any real trend. DLA contract obligations declined by 3 percent in 2016, largely due to a 37 percent decline in contract obligations for Fuels that CSIS believes is similarly related to the timing of large contracts. And contract obligations by Other DoD contracting entities were virtually stable.

2.2. Trough in Development Pipeline for Major Weapons Systems Continues into Seventh Year

In Chapter 4, CSIS discusses the massive decline in DoD R&D contract obligations during the budget drawdown. But as Figure 2-2 shows, that decline appears to have reached its floor in 2016:

---

[9] As discussed in detail in Chapter 7 of this report, all F-35 contract obligations are currently entered into FPDS as Navy contracts, so this reflects changes in obligations for the entirety of the F-35 program, not just the Navy/Marine Corps portions.

Figure 2-2: Defense Contract Obligations by Area, 2009–2016

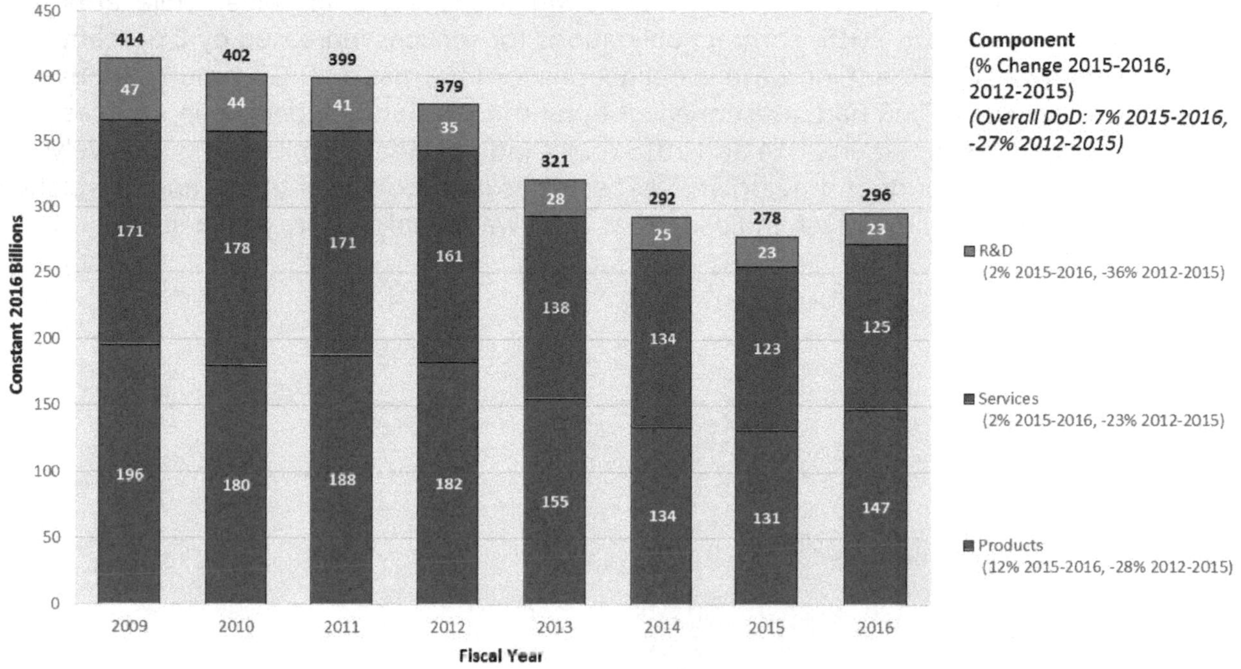
Constant 2016 Billions
Fiscal Year
2009
2010
2011
2012
2013
2014
2015
2016
414
402
399
379
321
292
278
296
Component
(% Change 2015-2016, 2012-2015)
(Overall DoD: 7% 2015-2016, -27% 2012-2015)
R&D
(2% 2015-2016, -36% 2012-2015)
Services
(2% 2015-2016, -23% 2012-2015)
Products
(12% 2015-2016, -28% 2012-2015)

Source: FPDS; CSIS analysis.

After declining by 36 percent between 2012 and 2015, overall DoD R&D contract obligations rose by 2 percent in 2016, driven by significant increases in obligations for both Applied Research (6.2) and Advanced Component Development & Prototypes (6.4).[10] The latter is particularly noteworthy, given the recent focus both within DoD and in Congress on promoting the use of prototyping in DoD development programs. At the same time, despite the slight increase in overall DoD R&D, contract obligations for System Development & Demonstration (6.5) continued their steep decline; after falling 57 percent between 2012 and 2015, System Development & Demonstration declined a further 15 percent in 2016, primarily within the Army and Navy. It is a measure of how deep this trough has run that, in 2016, DoD obligated only $500 million more for System Development & Demonstration contracts than it did for Clothing & Subsistence contracts.

This continued decline in contract obligations is reflective of the continued trough in DoD's development pipeline for major weapons systems, particularly within the Army. While the Navy's Columbia-class ballistic missile submarine program has just entered Milestone B, and the Air Force has funding for its B-21 bomber program beginning to ramp up, the Army faces continued budget pressures and uncertainty about future roles and missions. Because of this, Army R&D is likely to remain in this trough for the foreseeable future.

---

[10] Careful readers may notice that the topline totals for DoD R&D prior to 2015 in the above figure are $1 billion–$2 billion lower than in previous charts in this report. In the course of re-downloading the back-years FPDS data, CSIS discovered that, sometime after February 2015, MDA reclassified major portions of its contracting portfolio from between 2000 and 2014, changing the categorization of over $21 billion of contract obligations over that period from R&D to products. CSIS is currently engaging with DoD officials to understand when and why this reclassification of back-year MDA contracting data occurred.

DoD contract obligations for products increased by 12 percent in 2016, driven by large increases in obligations for Aircraft, Launchers & Munitions, and Ships. Meanwhile, after declining by 8 percent in 2015, contract obligations for services increased by 2 percent in 2016, driven by a 10 percent increase in obligations for Information & Communications Technology services. CSIS had questioned whether the accelerating decline in services contract obligations between 2014 and 2015 might indicate that despite being relatively preserved during the budget drawdown, DoD services contracts might be in line for relative decline; the 2016 data, however, indicates that 2015 was simply a one-year decline.

3. DoD Contract Spending in a Budgetary Context

Before looking at specific questions on the state of defense acquisition, it is helpful to understand how DoD's contract spending fits into the larger budgetary picture. To allow for a like-to-like comparison, CSIS compared DoD contract obligations to total DoD obligations, shown in Figure 3-1. These totals include contract spending associated with foreign military sales through the Defense Security Cooperation Agency (DSCA), and therefore may not match those reported by other sources.[11]

Figure 3-1: Defense Contract Obligations vs. Total Net Obligations, 2008–2015[12]

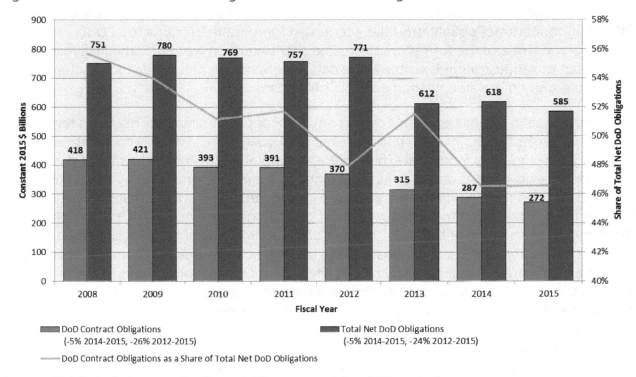

DoD Contract Obligations
(-5% 2014-2015, -26% 2012-2015)

Total Net DoD Obligations
(-5% 2014-2015, -24% 2012-2015)

——— DoD Contract Obligations as a Share of Total Net DoD Obligations

Source: FPDS; DoD Comptroller Financial Summary tables; CSIS analysis.

In this chart, the blue bar shows total DoD contract obligations, while the red bar shows total net DoD obligations, which includes both contract and noncontract obligations. The green line, which uses the secondary axis on the right, shows contract obligations as a share of total net obligations.

Overall DoD contract obligations peaked at $421 billion in 2009 following a steady increase throughout the 2000s. Total net DoD obligations similarly peaked in 2009, at $780 billion.

---

[11] See Appendix A, Methodology, for details on how CSIS calculates total net obligations.
[12] This chart includes data starting in 2008 because, while the required DoD comptroller data for overall DoD is available starting in FY2003, DSCA data are only broken out separately starting in FY2008.

Between 2009 and 2015, overall DoD contract obligations declined by 35 percent. Despite the declines in both DoD contract and net DoD obligations, the result of steady year-to-year declines initially associated with decreases in the Overseas Contingency Operations account, and accelerated in 2013 due to the impact of sequestration and the imposition of lower Budget Control Act caps, however, the shape of these declines exhibits substantial differences between contract obligations and noncontract obligations.

Total net DoD obligations were largely steady from 2009–2012 before declining by 21 percent in 2013 alone, notably more steeply than the 15 percent decline for contract obligations in that year. For the entire 2009–2015 period, however, total net DoD obligations have declined by 25 percent, notably less steeply than for contract obligations. As a share of total net DoD obligations, contract obligations declined from 56 percent in 2008 to 48 percent in 2012, then rose to 52 percent in 2013, and have remained at 47 percent in 2014 and 2015.

Contract obligations accounted for more than half of total net DoD obligations from 2008 to 2011, but noncontract obligations have accounted for more than half of total DoD obligations in three of the four years since. The only exception was 2013, the first year where the impact of sequestration is visible in the data, where noncontract obligations actually declined slightly more steeply than contract obligations.

The divergence in trends between contract and noncontract obligations may be a temporary phenomenon, however. In 2015, both total net DoD obligations and DoD contract obligations declined by 5 percent. This represents a notable slowing of the rate of decline for DoD contract obligations and may indicate that the decline in contract obligations has reached its floor. If nothing else, the data may indicate that trends in contract obligations may be more closely tied to trends in total net obligations in the near term, rather than declining independently of trends in total net obligations. And while CSIS is still processing the data on total net DoD obligations in 2016, the significant increase in contract obligations discussed in Chapter 2 make it likely that the decline of contract obligations as a share of total DoD obligations is at an end. And given the defense spending increase Congress provided in 2016 and the gradually increasing course of defense spending laid out in the last several years of the Budget Control Act through 2021, the future looks reasonably positive for DoD contracting. All of this, of course, is subject to adherence to the budget caps put in place by the Budget Control Act. While a budget agreement to significantly modify those caps has been elusive, the possibility remains that they could be adjusted, driving more substantial changes in DoD contract obligations. Additionally, with a new administration that campaigned on increasing defense budgets, and a Congress that seems amenable to going along with that policy, it seems likely that there will be more money available to DoD for contracting in the coming years.

4. What Is DoD Buying?

When most people think of DoD contracting, they think of contracts for major defense acquisition programs (MDAPs): large, expensive platforms and weapons systems that cost hundreds of millions, if not billions, to develop and procure. In truth, however, contract obligations related to MDAPs have only accounted for roughly a quarter of DoD contract obligations in recent years. DoD contracts for products range from aircraft carriers and satellites to basic commodities and commercial goods; for services ranging from highly specialized engineering and technical advising to landscaping and janitorial work; and for research and development (R&D) ranging from basic, fundamental research to highly focused development on specific platforms.

This chapter focuses on exploring the full range of what DoD contracts for, and examining the disparate impacts of the budget drawdown on different parts of DoD's contracting portfolio. The first section of this chapter looks at both DoD's policy efforts to promote innovation and the impact of the budget drawdown on DoD's R&D contracting portfolio. The second section divides DoD contracts into what CSIS has termed "platform portfolios," examining trends in contracts related to different types of major DoD platforms. And the third section examines DoD contracting through the prism of the budget process, by examining changes to how contract obligations are funded out of particular budget accounts.

As described in Chapter 1, overall DoD contract obligations have declined by 35 percent since their peak in 2009. But that decline has not been evenly distributed across the breadth of DoD contract obligations. To examine trends within DoD's contracting portfolio, CSIS breaks down contract obligations into three contracting market areas: products, services, and R&D, as seen in Figure 4-1:

Figure 4-1: Overall Defense Contract Obligations by Area, 2000–2015

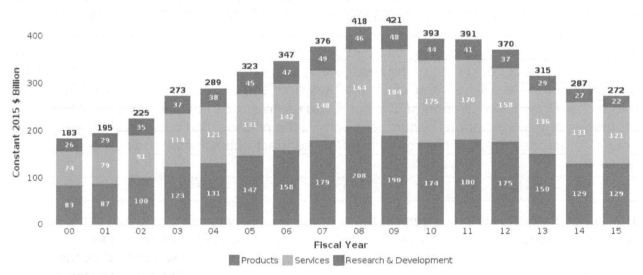

Source: FPDS; CSIS analysis.

Since 2009, DoD contract obligations for products (-32 percent) and services (-34 percent) have declined roughly in parallel to the overall decline in DoD contract obligations between 2009 and 2015. Products contract obligations were actually fairly stable between 2009 and 2012 before declining strongly since, while services have been steadily declining since 2009, with a spike in the rate of decline in 2013. Between 2012 and 2015, as overall DoD contract obligations fell by 26 percent, both products (-26 percent) and services (-23 percent) again declined roughly in parallel to the overall rate of decline for DoD contract obligations. That services have declined slightly less steeply than overall DoD trends is notable, because there was a great deal of speculation throughout the budget drawdown that services would be disproportionately targeted for savings.

In 2015, as overall DoD contract obligations declined by 5 percent, DoD products contract obligations were stable (0 percent), after suffering a 14 percent decline in 2014. Year-to-year trends in products contract obligations are highly sensitive to the timing of large contracts for production of major weapons systems such as the F-35, however, and should be evaluated in that context, though with purchases of F-35s likely to accelerate in the near future, these large contracts may be a continuing source of stability within the DoD products contract portfolio. DoD contract obligations for services fell by 8 percent, more steeply than for DoD contracts overall, after declining by only 4 percent in 2015. This decline is broad based, with four of the five categories of services contracts (excepting Professional, Administrative, and Management Support services) declining more steeply in 2015 than in 2014, with an average difference of -13 percentage points. This accelerating decline in DoD services contracts, as the overall decline in DoD contracts seems to be close to reaching its floor, appeared to indicate that services, or at least some service categories, were poised to face disproportionate declines in the coming years. But as the FY2016 data discussed in Chapter 2 showed, this accelerating decline appears to have been a one-year phenomenon, rather than the start of a larger trend.

The outlier among the contracting market areas is R&D—since 2009, DoD R&D contracting obligations have declined by more than half (-53 percent), over half-again as steeply as overall DoD contract obligations. Since 2012, R&D contract obligations have declined by 39 percent, also half again as steeply as the overall rate of decline for DoD contracts.

In 2015, R&D contract obligations declined by 17 percent, over three times the rate of overall DoD contracts, after declining slightly less steeply (-7 percent) than DoD contracts overall (-9 percent) in 2014. As will be discussed in the "Research and Development Contracting during the Budget Drawdown" section of Chapter 4, the relative dearth of new development programs for major weapons systems currently in the pipeline, particularly for the Army, is likely to preclude any significant growth in DoD contract obligations in the near term.

4.1. Innovation, R&D, and Technological Superiority

The search for innovation has been at the heart of key policy issues for DoD over the past year. The Third Offset Strategy was launched as a way to focus R&D efforts on key technologies and capabilities that DoD feels will shape the future of warfare, while the Defense Innovation Unit Experimental (DIUx) initiative was launched as a way to identify innovation in the commercial sector, and to find ways to bring that innovation into the defense realm. Both of these initiatives have faced questions and challenges over the past year, and their respective futures are uncertain, given that a new administration, with its own priorities and policy goals, is incoming. These efforts, or ones like them, are particularly important now, as the huge decline in R&D contract obligations over the course of the budget drawdown has led to a significant trough in the development pipeline for new capabilities. With declining funding for contracted R&D, DoD needs to better focus its efforts on capabilities that will drive future technological superiority, and needs to expand its search parameters outside of the traditional defense R&D industrial base in order to ensure that those capabilities are developed and fielded in a timely and cost-effective manner.

4.1.1. Defense Innovation Initiative—"Third Offset Strategy" Update

A key element of the broader defense innovation push, the Third Offset Strategy remained a top DoD priority in 2016. Championed by Deputy Secretary of Defense Bob Work, the Third Offset Strategy is being pursed to counter great powers (China and Russia) and ensure continued U.S. military technological superiority.

The Third Offset Strategy, originally launched in 2014, aimed at ensuring America's continued technological military dominance remains a top DoD priority. After a period of ambiguity, Deputy Secretary of Defense Robert Work announced in November 2015 that Third Offset's "big idea" was human-machine collaboration and combat teaming.[13] Since November 2015, there have been three developments: investments in the FY2017 President's Budget (PB), congressional response, and the declassification of the Strategic Capabilities Office.

PB 17 Investments

A key indicator of the ultimate success of Third Offset will be transitioning capabilities and technologies from ideas and into the budget and ultimately programs of record. With the first opportunity to do so being the PB2017 request, there were large expectations.

Prior to its unveiling, Deputy Secretary Work indicated that there would be approximately $12 billion–$14 billion invested over the Future Years Defense Program (FYDP) in the 2017 President's Budget (PB).[14]

Released publicly in February 2016, the President's FY2017 budget included $18 billion over the course of the FYDP for Third Offset-related investments—$3.6 billion of which being spent in FY2017. Within that $18 billion, DoD elected to make smaller bets over targeted larger investments in specific capabilities. Over the FYDP, DoD invested $3 billion on anti-access technologies, undersea warfare capabilities, and human-machine collaboration; $1.7

---

[13] For a discussion of the history of the Third Offset Strategy, see Jesse Ellman et al., "Defense Acquisition Trends, 2015: Acquisition in the Era of Budgetary Constraints," Center for Strategic and International Studies, January 25, 2016, 7–10, https://csis-prod.s3.amazonaws.com/s3fs-public/legacy_files/files/publication/160126_Ellman_DefenseAcquisitionTrends_Web.pdf.

[14] Aaron Mehta, "Work Outlines Key Steps in Third Offset Tech Development," *DefenseNews*, December 14, 2015, http://www.defensenews.com/story/defense/innovation/2015/12/14/work-third-offset-tech-development-pentagon-russia/77283732/

billion on cyber and electronic warfare technology; and $500 million on precision-guided munitions and war-gaming.[15]

Hill Response to the Third Offset Strategy

Congressional response to Third Offset has been largely positive, but not without some reservation.[16] Many members of Congress believe modernizing DoD requires further reforms. Representative Mac Thornberry, chairman of the House Armed Services Committee, while applauding DoD's move to further innovate and advance, warned that much more change is needed to maintain military dominance than new technological innovations. Thornberry lists cyber, nuclear deterrence, and special operations as key fields for continued development if America's military is to remain the finest military in the world. According to Rep. Thornberry, new technologies will not solve the insecurities of cyber, for that the department needs to focus on "people, organization, rules of engagement in that domain to try to make sure we close the gap between the threat and the policies we now have to deploy." Rep. Thornberry also acknowledges that nuclear deterrence "may seem a little bit odd" but North Korea has conducted two nuclear tests thus far in 2016, and five since 2006.[17] Proliferation is one issue, but the decaying state of the U.S. nuclear fleet and its infrastructure is a huger concern for Rep. Thornberry. Lastly, Thornberry has emphasized the importance and utility of our special forces. However, these forces have been extensively used in the recent decades to the point of potentially being overused and therefore losing utility.

Strategic Capabilities Office

Part of the broader Defense Innovation Initiative, earlier this year Secretary Carter publicly revealed the mission of the previously classified Strategic Capabilities Office (SCO). Established in 2012 by then-deputy secretary of defense Ash Carter, SCO works to ensure America's lead in military technological capabilities by mixing and merging technologies across multiple platforms and services.[18] Led by Dr. Will Roper, SCO works to "reimagine existing DoD, intelligence or commercial capabilities" instead of focusing on creating wholly new technologies.[19]

> In SCO's relatively short lifespan, the office has had significant successes that include, but are not limited to, the hyper velocity projectile, arsenal plane, and the Army's Tactical Missile System (ATACMS). The hyper velocity projectile took the projectile initially developed for the Navy's electromagnetic railgun and employed it on existing

---

[15] Aaron Mehta, "Defense Department Budget: $18B Over FYDP for Third Offset," *DefenseNews*, February 9, 2016, http://www.defensenews.com/story/defense/policy-budget/budget/2016/02/09/third-offset-fy17-budget-pentagon-budget/80072048/

[16] Jen Judson, "Thornberry to Prioritize Third Offset, Cyber, Nuke Modernization, Special Ops," *DefenseNews*, January 13, 2016, http://www.defensenews.com/story/breaking-news/2016/01/13/thornberry-prioritize-third-offset-cyber-nuke-modernization-special-ops/78753522/.

[17] Sharon Squassoni and Amelia Armitage, "Fifth DPRK Nuclear Test," *Center for Strategic and International Studies*, September 9, 2016, https://www.csis.org/analysis/fifth-dprk-nuclear-test.

[18] Sam LaGrone, "Little Known Pentagon Office Key to U.S. Military Competition with China, Russia," *USNI News*, February 2, 2016, https://news.usni.org/2016/02/02/little-known-pentagon-office-key-to-u-s-military-competition-with-china-russia

[19] Ibid.

Naval five-inch guns to use as potential cruise and ballistic missile defense. The SCO-built arsenal plane uses one of the US' oldest air platforms to create an aerial launchpad for a variety of conventional payloads. This aerial launchpad of sorts links with fifth-generation aircraft that will serve as "forward sensor and targeting nodes" which allows the military to repurpose two current technologies into a new capability.[20] SCO work on the ATACMS focused on allowing the surface-to-surface missile now target moving targets on both land and at sea.[21]

An indicator of the importance and success of SCO is the growth of its budget. Since its creation in 2012, the SCO budget has grown considerably as reliance on and confidence in SCO's mission and breakthroughs increased within the Pentagon. In 2014, the SCO budget totaled just $125 million. In 2015, the SCO budget increased to $175 million before substantially increasing to approximately $530 million in 2016.[22] In PB17, the SCO budget request totaled $902 million.[23] The significant growth in the budgets reflects a significant amount of "buy-in" from the military services, particularly the Navy and Army, which is another key metric of success for SCO.

SCO and the Third Offset?

Created prior to the Defense Innovation Initiative, the SCO quickly became a critical component of the ongoing innovation effort. SCO does not seek to replace Third Offset, but instead complement the effort. Whereas Third Offset is focused on developed next-generation technologies, SCO focuses on improving current-generation technologies. By improving current-generation technologies, SCO is seeking to buy DoD 10 to 15 years so that the next-generation Third Offset technologies can mature.[24]

Watch to Watch For: Transition to the New Administration

Going forward, a key question is whether the Defense Innovation Initiative, and its components, survive the transition to the new Trump administration. Throughout the campaign, the president-elect and his surrogates made rebuilding the military a focus of the campaign, but made little reference to military innovation in particular. Instead, the campaign focused on rebuilding capacity through increases to Army active duty force structure, Naval warship counts, and age of Air Force fleet. The selection of retired Gen Mattis, U.S. Marine Corps, provides little guidance in either direction regarding the long-term future of these efforts. Throughout his active duty career, Gen Mattis largely focused on operational

[20] Colin Clark and Sydney Freedberg, "Robot Boats, Smart Guns & Super B-52s: Carter's Strategic Capabilities Office," *Breaking Defense*, February 5, 2016, http://breakingdefense.com/2016/02/carters-strategic-capabilities-office-arsenal-plane-missile-defense-gun/.

[21] Aaron Mehta, "Anti-Naval ATACMS, 'Big' Swarming Breakthroughs from Strategic Capabilities Office," *DefenseNews*, October 28, 2016, http://www.defensenews.com/articles/anti-naval-atacms-big-swarming-breakthroughs-from-strategic-capabilities-office

[22] Clark and Freedberg, "Robot Boats, Smart Guns & Super B-52s: Carter's Strategic Capabilities Office."

[23] Aaron Mehta, "Defense Department Budget: $18B Over FYDP for Third Offset," *DefenseNews,* February 9, 2016, http://www.defensenews.com/story/defense/policy-budget/budget/2016/02/09/third-offset-fy17-budget-pentagon-budget/80072048/

[24] Sydney J. Freedberg Jr., "Strategic Capabilities Office Is 'Buying Time' for Offset: William Roper," *Breaking Defense*, July 18, 2016, http://breakingdefense.com/2016/07/strategic-capabilities-office-is-buying-time-william-roper/

activities, not defense innovation. However, the recent decision by the transition team to temporarily retain Deputy Secretary Work suggests that the Third Offset initiative may not be dead with the new administration. Whomever is eventually selected to replace Deputy Secretary Work will be an indicator about the long-term future of the Defense Innovation Initiative.

Within DoD, senior leaders are undertaking steps to preserve the Third Offset Strategy. For example, Deputy Secretary Bob Work has outlined three steps that the department is taking to preserve the Third Offset Strategy in the Trump administration. [25] First, DoD is giving the administration options from which to select. For example, within programs, the incoming DoD leadership can select from multiple choices. Second, DoD is working to continue to build buy-in with Congress. As discussed previously, DoD has already found a warm, but cautious, constituency on the Hill. Finally, DoD is working to build buy-in from uniformed military leaders who will outlast political appointees. Within the uniformed military, there is already high-level buy-in from vice chairman of the Joint Chiefs, Gen Paul Selva. [26] Within the services, the service chiefs have spoken highly of the Third Offset Strategy, but it remains to be seen what level of buy-in exists within the current political leadership. [27]

Defense Innovation Unit Experimental 2.0

DIUx is the effort by DoD to bridge relations between the Pentagon and commercially focused innovators in places like Silicon Valley in order to create new partnerships and encourage more cooperation. DIUx opened its doors in August 2015 and since then the organization "has made connections with more than 500 entrepreneurs and firms." [28] Additionally, DIUx has hosted many forums that connected senior DoD officials with innovators in Silicon Valley. DIUx has also been instrumental in funding almost "two dozen technology projects—spanning everything from wind-powered drones to data analytic tools." [29]

Originally instructed to "serve as the hub for the department's core initiative to increase DoD's communication with, knowledge of, and access to innovating, leading-edge technologies from high-tech startups and entrepreneurs," [30] DIUx has quickly gone through a few updates to its mission over the past year. In fact, in May 2016, Secretary Carter announced DIUx 2.0, the next version of the organization that modified DIUx's management structure to make it more similar to innovation facilitators in Silicon Valley such as venture

---

[25] Aaron Mehta, "Pentagon No. 2: How to Keep Third Offset Going in the Next Administration," *DefenseNews*, May 2, 2016, http://www.defensenews.com/story/defense-news/2016/05/02/pentagon-no-2-how-keep-third-offset-going-next-admininistration/83851204/.

[26] Ibid.

[27] Sebastian Sprenger, "Army takes wait-and-see stance on Pentagon's 'Third Offset' strategy," *Inside Defense*, December 18, 2015, https://insidedefense.com/inside-army/army-takes-wait-and-see-stance-pentagons-third-offset-strategy.

[28] U.S. Department of Defense, "Remarks Announcing DIUx 2.0," May 11, 2016, http://www.defense.gov/news/speeches/speech-view/article/757539/remarks-announcing-diux-20.

[29] Ibid.

[30] U.S. Department of Defense, "Readout of Deputy Secretary of Defense Bob Work and Undersecretary of Defense for Acquisition, Technology, and Logistics Frank Kendall Visit to Defense Innovation Unit Experimental," August 5, 2015, http://www.defense.gov/news/news-releases/news-release-view/article/612824/readout-of-deputy-secretary-of-defense-bob-work-and-undersecretary-of-defense-f.

capital partnerships. DIUx now has a managing partner rather than a director and began reporting directly to the secretary. The largest alteration is the scaling-up of the organization by expanding it to other locations around the country. The second office, located outside Silicon Valley, was formally opened in July 2016 in Boston, which has been designated as another hub of innovation by the Department.[31] In September 2016, DIUx announced a third office in Austin, Texas. Along with this announcement, Secretary Carter revealed that the DIUx will have about $65 million in contracts to award in the near future.[32] Other innovation hubs where DIUx may locate branches include but are not limited to Silicon Alley in New York City; Atlanta, Georgia; the North Carolina Research Triangle; the Puget Sound region in Washington; and Northern Virginia.[33]

Also part of DIUx 2.0, Secretary Carter announced that the Defense Department would be requesting an additional $30 million in direct funding for the organization. This amount is not the ceiling of available funding, though, as this number does not include coinvested funds that can be provided by the services. Secretary Carter also promised that DIUx would use this funding to host merit-based prize competitions, encourage and fund incubator partnerships, and work to target R&D efforts that fund promising technologies. DIUX will expand its focus to help connect nontraditional industry to DoD's rapid acquisition requirements. In addition, DIUx is utilizing the Commercial Solutions Opening (CSO), a new contracting tool developed by Army Contracting Command that allows commercial companies to submit proposals in response to DIUx's call for new technologies and get a quick answer.[34] DIUx will now also be organized into three units: the Engagement Team, the Foundry Team, and the Venture Team. These three groups have been dedicated to separate parts of DIUx's mission.[35] Carter stated that the main goal of the Engagement Team would be to introduce "the military to entrepreneurs," but more importantly to introduce "entrepreneurs to military problems." The Foundry Team will work with maturing technology or technology that needs to be altered so that DoD can use it. The final team, and the largest, the Venture Team, will look at how applicable commercial products and technology can be rapidly adapted to use within the U.S. military.[36]

The rapid pace of change at DIUx is one indication of the priority placed on innovation in 2016 and of the willingness of DoD leadership to rapidly adapt and adjust its approach based on feedback and experience. However, the speed of development of DIUx has also led to concerns in Congress. The final version of the National Defense Authorization Act for Fiscal Year 2017 (Section 222) limits the expenditure of both operation and maintenance funding

---

[31] U.S. Department of Defense, "Carter Speaks at Boston Defense Innovation Unit Experimental," Videos, July 26, 2016, http://www.defense.gov/video?videoid=476363#.v5fs57lvqeg.

[32] Aaron Mehta, "Carter: $65 Million in DIUx Contracts in Pipeline | DefenseNews," *DefenseNews*, September 14, 2016, http://www.defensenews.com/articles/carter-65-million-in-diux-contracts-in-pipeline.

[33] Stephanie Walden, "Beyond Silicon Valley: Top Emerging Startup Markets in the U.S.," *Mashable*, August 19, 2015, http://mashable.com/2015/08/19/top-new-cities-startup-markets/#m29bba.w75qu.

[34] Tony Bertuca, "Carter opens DIUx-Boston and reveals latest tech sector outreach efforts," *Inside Defense*, vol. 32, no. 30, July 28, 2016, 4.

[35] Cheryl Pellerin, "Carter Opens Second DoD Innovation Hub in Boston.," U.S. Department of Defense, July 26, 2016, http://www.defense.gov/news-article-view/article/858413/carter-opens-second-dod-innovation-hub-in-boston.

[36] Aaron Mehta, "Carter Announces Changes, Contracts for DUIx," *DefenseNews*, July 26, 2016, http://www.defensenews.com/story/defense/innovation/2016/07/26/carter-announces-structural-changes-contracts-diux/87572286/.

(to no more than 75 percent of the annual amount) and R&D funding (to no more than 25 percent of the annual amount) for DIUx until an extensive report explaining the plans for DIUx and justifying its expansion is submitted to Congress. This provision represents a shot across the bow for DoD, raising concerns about the exact nature of DIUx's mission, how it fits into the larger scheme of the acquisition system, as well as its rapid growth. It remains to be seen if these concerns represent an existential challenge to DIUx as it attempts to make the transition to the new administration.

Defense Innovation Advisory Board

In a related development, Secretary Carter established a new Federal Advisory Committee Act entity to advise DoD on accessing and implementing innovation known as the Defense Innovation Advisory Board. The board includes a roster of influential Silicon Valley technologists, along with other experts, including Amazon's Jeff Bezos, Code for America founder Jennifer Pahlka, and scientist Neil deGrasse Tyson. The board's purpose is to identify a range of innovative private-sector practices and technological solutions that DoD can adapt. The board will likely function like other FACA entities such as the Defense Science Board and the Defense Business Board, advising DoD on questions posed by DoD leadership. The establishment of the Defense Innovation Advisory Board attempts to ensure that DIUx, or at least its mandate, will live on past Secretary Carter's leadership of the Pentagon.

Changes to DoD Organization in 2017 National Defense Authorization Act

The National Defense Authorization Act for Fiscal Year 2017 includes substantial changes intended to ensure that innovation is established as a paramount institutional priority for DoD through the establishment of a new under secretary of defense for research and engineering, through the restricting of how technology development is organized and funded within DoD, and through changes in how DoD's investment in technology is implemented into contracts with industry. These topics are explored in depth in Chapter 3 as part of the discussion of acquisition reform.

4.1.2. Research and Development Contracting during the Budget Drawdown[37]

Technological superiority has been a central pillar of U.S. strategy in the post–World War II era. It has allowed the United States to deter, and when necessary defeat, numerically superior forces of potential or actual adversaries. But with other nations building their capabilities and infrastructure at a rapid pace, it is not safe or wise to assume that U.S. technological superiority is a foregone conclusion. Furthermore, as the current budget drawdown has progressed, numerous analysts and policymakers have expressed concern regarding the ability of the United States to retain technological superiority, particularly given how research and development (R&D) contracting has been broadly understood to be in serious decline. Broadly speaking, the stated concerns can be summarized as a fear that the

---

[37] This section is adapted from CSIS's report: "Federal Research and Development Contract Trends and the Supporting Industrial Base, 2000–2015" (https://csis-prod.s3.amazonaws.com/s3fs-public/publication/160914_Ellman_FederalRDContractTrends_Web.pdf), which was performed with support from the Naval Postgraduate School's Acquisition Research Program.

R&D necessary to drive future technological breakthroughs, in either the defense or civilian realms, would be jeopardized and would be particularly damaged if agencies disproportionately sacrificed longer-term R&D spending in order to preserve current programs and activities.

To analyze trends within DoD's R&D contracting portfolio, CSIS has developed a methodology to categorize R&D contracts by stage of R&D, using a categorization schema that roughly corresponds to the commonly used DoD R&D Budget Activity Codes (BACs):[38]

- Basic Research (6.1)

- Applied Research (6.2)

- Advanced Technology Development (ATD) (6.3)

- Advanced Component Development & Prototypes (ACD&P) (6.4)

- System Development & Demonstration (SD&D) (6.5)

- Operational Systems Development (6.7)

- Operation of Government R&D Facilities (GOCO)[39]

Since 2009, DoD R&D contract obligations have declined by 53 percent, notably faster than the 35 percent decline in overall DoD contract obligations over this same period. As a share of overall DoD contract obligations, R&D declined from 11 percent in 2009 to 8 percent in 2015, the lowest share seen in the 2000–2015 period. Figure 4-2 shows the breakdown of DoD R&D contract obligations by stage of R&D:

---

[38] CSIS does not include contracts for R&D Management Support (6.6) in this analysis.
[39] Though not classified as R&D in FPDS, CSIS now includes the codes for management/operation of federal R&D facilities in its R&D category, as a significant amount of R&D activity in the broader federal government is structured in this manner.

Figure 4-2: DoD R&D Contract Obligations by Stage of R&D, 2000–2015[40]

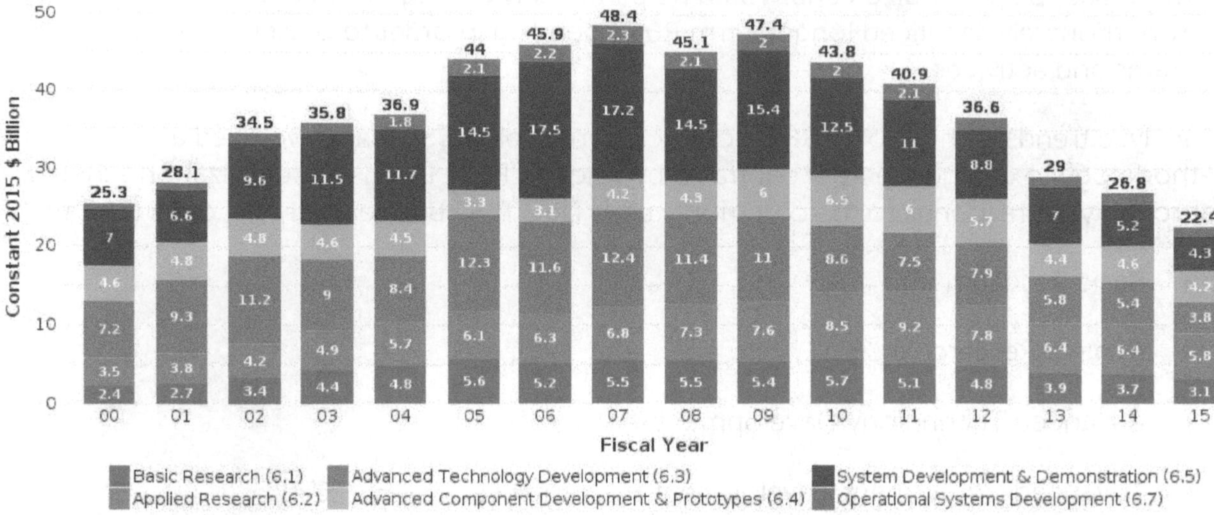

Source: FPDS; CSIS analysis.

Since 2009, as overall DoD R&D contract obligations declined by 53 percent, obligations for Applied Research declined by less than half that rate (-23 percent),[41] while obligations for Basic Research declined by only 42 percent. As a share of DoD R&D contract obligations, the two seed-corn categories rose from 27 percent in 2009 to 40 percent in 2015, the highest share in the 2000–2015 period. Basic Research contract obligations have declined at a rate that more closely parallels the overall decline in DoD R&D contract obligations since 2012, but Applied Research obligations have continued to be relatively preserved (-25 percent decline since 2012, compared to -39 percent for overall DoD R&D).

Contract obligations for ACD&P (-31 percent) and Operational Systems Development (45 percent) have similarly been relatively preserved since 2009, though the latter declined by 26 percent in 2015, nearly half again as steeply as overall DoD R&D (-17 percent). But ATD (-65 percent) and SD&D (-72 percent) have seen massive declines in recent years. The declines in ATD and SD&D accounted for nearly three-quarters of the total decline in DoD R&D contract obligations during the current drawdown.

DoD contract obligations for SD&D (-18 percent) and Basic Research (-14 percent) fell roughly in parallel to overall DoD R&D in 2015, but obligations for ATD fell notably more steeply (- 29 percent,) while obligations for Applied Research (-9 percent) declined at roughly half the rate of overall DoD R&D.

The enormous decline in SD&D is particularly telling and speaks to the larger trend in DoD R&D contracting—over the past several years, as R&D programs related to MDAPs have either been canceled or matured into production, DoD has been largely unable to start and sustain new development programs, due either to budgetary pressures or to programmatic

---

[40] The massive reclassification of back-years MDA R&D contract obligations discussed in Chapter 2, such that they are no long categorized as R&D, has altered the magnitude of the changes analyzed in this section, but has not significantly altered the overall trend line.

[41] DoD contract obligations for Applied Research actually saw a notable spike between 2009 and 2011, due primarily to a one-year spike for space-related R&D, but obligations returned to prior levels in 2012.

difficulties. The decline in R&D contract obligations during the budget drawdown thus appears to reflect a six-year trough in the pipeline of new major weapons systems; as the FY2016 data discussed in Chapter 2 shows, this trough has extended into a seventh year. The dimensions of this trough will be discussed further in the sections that follow.

Trends in defense R&D contracting are not uniform across the military services and other major R&D contracting components within DoD, as can be seen in Figure 4-3:

Figure 4-3: DoD R&D Contract Obligations by Component, 2000–2015

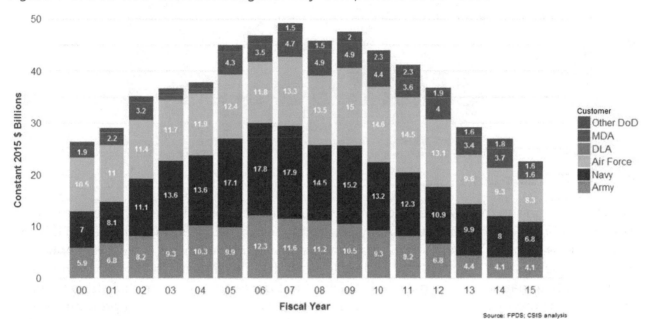

Source: FPDS; CSIS analysis.

The following sections will examine trends in R&D contracting within the three military services, plus the Missile Defense Agency (MDA), in greater detail.

Figure 4-4: Army R&D Contract Obligations by Stage of R&D, 2000–2015

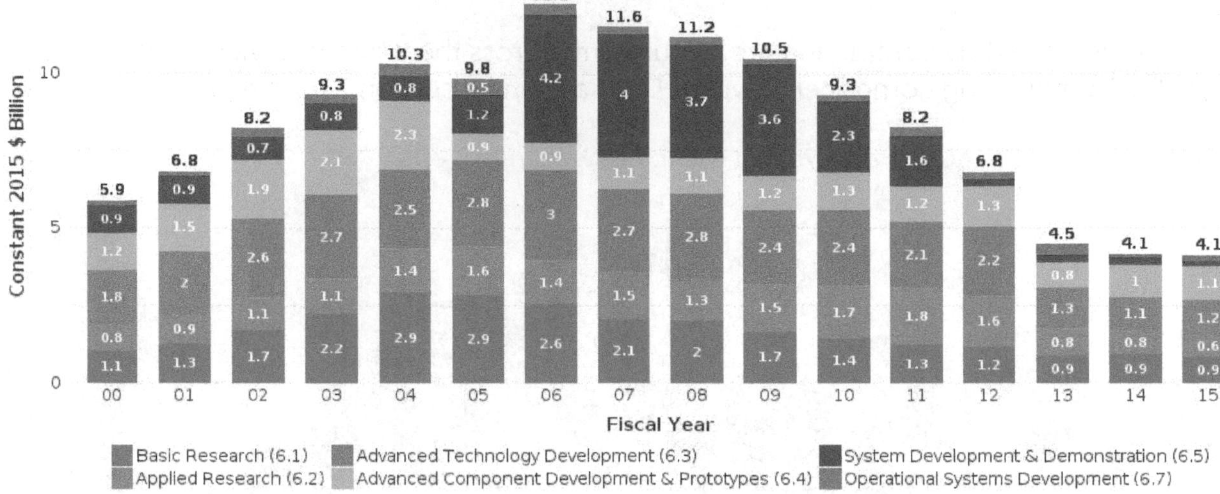

Constant 2015 $ Billion

Fiscal Year	00	01	02	03	04	05	06	07	08	09	10	11	12	13	14	15
Total	5.9	6.8	8.2	9.3	10.3	9.8	12.3	11.6	11.2	10.5	9.3	8.2	6.8	4.5	4.1	4.1
System Development & Demonstration (6.5)	0.9	0.9	0.7	0.8	0.8	0.5	4.2	4	3.7	3.6	2.3	1.6				
Advanced Component Development & Prototypes (6.4)	1.2	1.5	1.9	2.1	2.3	1.2	0.9	1.1	1.1	1.2	1.3	1.2	1.3	0.8	1	1.1
Advanced Technology Development (6.3)	1.8	2	2.6	2.7	2.5	2.8	3	2.7	2.8	2.4	2.4	2.1	2.2	1.3	1.1	1.2
Applied Research (6.2)	0.8	0.9	1.1	1.1	1.4	1.6	1.4	1.5	1.3	1.5	1.7	1.8	1.6	0.8	0.8	0.6
Basic Research (6.1)	1.1	1.3	1.7	2.2	2.9	2.9	2.6	2.1	2	1.7	1.4	1.3	1.2	0.9	0.9	0.9

Legend:
- Basic Research (6.1)
- Applied Research (6.2)
- Advanced Technology Development (6.3)
- Advanced Component Development & Prototypes (6.4)
- System Development & Demonstration (6.5)
- Operational Systems Development (6.7)

Source: FPDS; CSIS analysis.

The key factor in the massive decline in Army R&D contract obligations (-61 percent since 2009, compared to -55 percent for Army contracts overall) has been the cancelation of the Army's Future Combat Systems (FCS) program. Nearly all the decline in Army R&D contract obligations between 2009 and 2012 is directly attributable to the cancelation and winding down of FCS. In particular, obligations for SD&D have declined by an incredible 95 percent since 2009, as the Army has struggled to start and sustain new development programs for major weapons systems in the wake of FCS's cancelation. The result of these struggles is the current, now seven-year trough in the Army's development pipeline for major weapons systems.

In terms of seed-corn R&D, the trend within the Army is mixed. While Basic Research (-49 percent) has been relatively preserved since 2009, Applied Research (-58 percent) has declined nearly as steeply as overall Army R&D. The decline in Applied Research was not consistent throughout the period, however; Army obligations for Applied Research actually grew between 2009 and 2011, before declining by nearly half in 2013 and falling by another 18 percent in 2015. In 2015, combined obligations for the two seed-corn categories are at their lowest level ($1.5 billion) in the 2000–2015 period.

In 2015, Army R&D contract obligations were relatively stable (-1 percent), indicating that the decline may have finally reached its floor. In addition to the aforementioned steep decline for Applied Research in 2015, obligations for SD&D also fell by 18 percent, though that only represents a drop from $220 million to $180 million. Meanwhile, Basic Research declined moderately (-7 percent), while both ATD (10 percent) and ACD&P (4 percent) rose moderately.

At present, the Army has no major ground vehicle development programs on the horizon and continues to face significant budgetary pressures. With the Army struggling to define the missions it expects to focus on in the future, as well as the capabilities it will need to perform those missions, the trough in the Army's development pipeline for major weapons systems seems likely to continue for the foreseeable future. This is particularly worrisome because defense R&D has historically seen a cycle where investments made in growth periods show results during subsequent drawdown periods. For the Army, this pattern appears to have been broken.

This interruption of the developmental pipeline does present an unusual opportunity for DoD, and particularly for the Army. As spending on war materiel continues to be replaced by funding for next-generation priorities, the Army has little to no developmental money already committed to projects. Thus, the Army has an opportunity to take a step back, draw lessons from the wars in Iraq and Afghanistan, evaluate potential future threats and missions, and determine their requirements and developmental priorities accordingly.

Navy—Huge Declines in Basic Research and Mid- to Late-Stage R&D as Columbia-Class Ramp-up Looms

Figure 4-5: Navy R&D Contract Obligations by Stage of R&D, 2000–2015

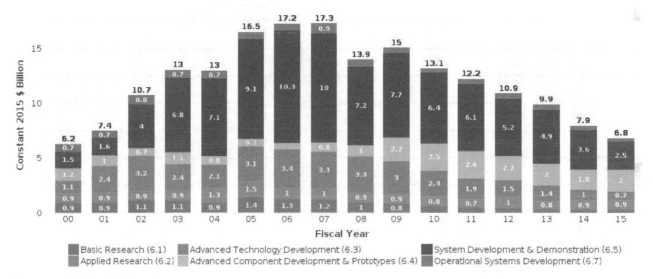

Source: FPDS; CSIS analysis.

While overall Navy contract obligations were relatively preserved (-20 percent) since 2009, Navy R&D contract obligations fell by 55 percent over that same period. As a share of overall Navy contract obligations, R&D fell from 14 percent in 2009 to 8 percent in 2014, which is the lowest share for the Navy in the 2000–2015 period.

Whereas obligations for Advanced Research have increased by 6 percent over the 2009–2015 period, obligations for Basic Research have declined by two-thirds since 2009. As with the Army, the Navy saw disproportionate declines in obligations for ATD (-75 percent) and SD&D (-67 percent). Unlike the Army, the Navy has major development programs in the pipeline, such as the Columbia-class ballistic missile submarine replacement. However, to preserve funding for current priorities, the Navy has been forced to push back the timelines for some of its efforts due to budgetary constraints, resulting in the ongoing trough in the Navy's development pipeline.

This trough is particularly visible in 2015—although overall Navy contract obligations were virtually stable (-1 percent), Navy R&D contract obligations declined by 14 percent. This decline was driven by a continued steep decline in both ATD (-22 percent) and SD&D (-30 percent). For SD&D, this represents the largest one-year decline in the period, and since 2014, Navy SD&D has declined by nearly half. In contrast, Navy obligations for Basic Research (3 percent) and Applied Research (4 percent) increased slightly in 2015; the increase in Basic Research is particularly notable, given the steep declines throughout the drawdown period, and represents the first increase to Navy Basic Research contract obligations since 2005.

For the Navy, then, there are two disparate trends within its R&D contracting portfolio. While the decline in Basic Research seems to have hit its floor and begun to rebound, the decline in mid- to late-stage R&D not only continues, but seems to have accelerated.

Air Force—B-21 to Reverse Decline in Mid- to Late-Stage R&D in Coming Years

Figure 4-6: Air Force R&D Contract Obligations by Stage of R&D, 2000–2015

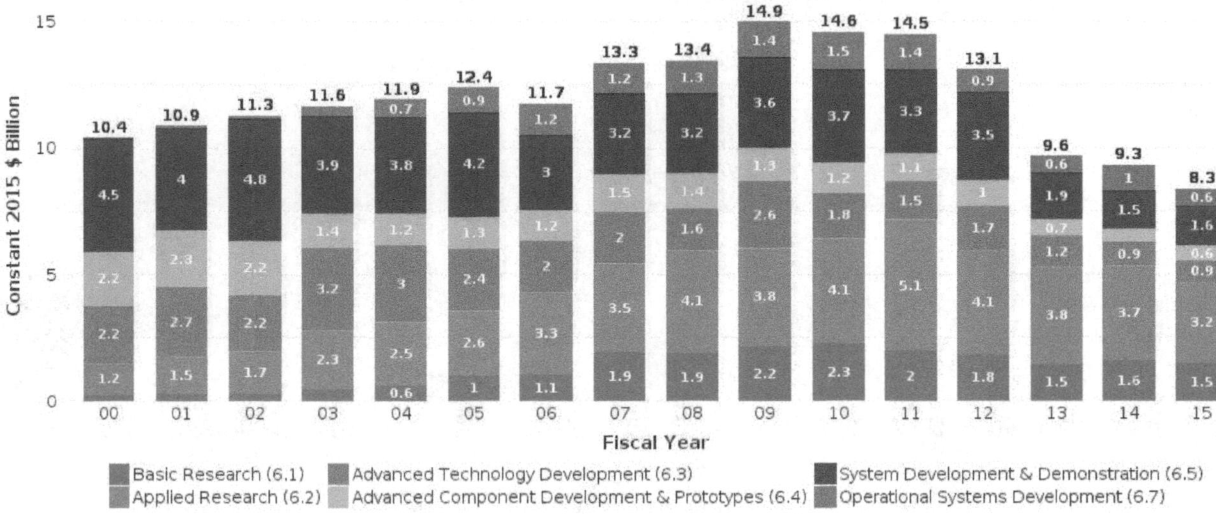

Source: FPDS; CSIS analysis.

As with the Navy, while overall Air Force contract obligations have been relatively preserved (-30 percent) between 2009 and 2015, R&D contract obligations within the Air Force declined more steeply (-44 percent) over that same period, though less steeply than DoD R&D contract obligations overall. Analogous to Army and Navy, Air Force contract obligations for Applied Research were relatively preserved since 2009 (-17 percent); unlike the Navy, Basic Research was also relatively preserved (-32 percent), and actually increased by 11 percent in 2014 before declining again in 2015. As a share of Air Force R&D contract obligations, seed-corn R&D rose from 41 percent in 2009 to 58 percent in 2014—the highest share in the 2000–2015 period, before falling back to 56 percent in 2015.

Both ATD (-67 percent) and SD&D (-57 percent) declined heavily, with the bulk of the declines coming in the wake of the main impact of sequestration between 2012 and 2013. Unlike both Army and Navy, however, Air Force contract obligations for ACD&P also declined heavily (-67 percent) since 2009.

In 2015, as overall Air Force contract obligations fell by 7 percent, Air Force R&D declined slightly more steeply (-10 percent). Both ACD&P (13 percent) and SD&D (5 percent) saw increases in 2015, while Applied Research fell by 14 percent, bringing Air Force Applied Research down to its lowest level since 2005. Interestingly, Air Force contract obligations for Operational Systems Development, which had fallen by nearly three-fifths between 2010 and 2013, rose by nearly two-thirds in 2014, before falling back to 2014 levels in 2015, indicating that the increase in 2014 was just a one-year spike.

The Air Force is also in the midst of a trough in their development pipeline for new major weapons systems, but with contracts recently awarded for major programs like the Long Range Strike Bomber and funding that's supposed to ramp up to significant levels over the next few years, the Air Force seems as if it will be the first of the military services to emerge from it.

Missile Defense Agency—Contract Obligations Decline by over Half in 2015[42]

Figure 4-7: MDA R&D Contract Obligations by Stage of R&D, 2000–2015

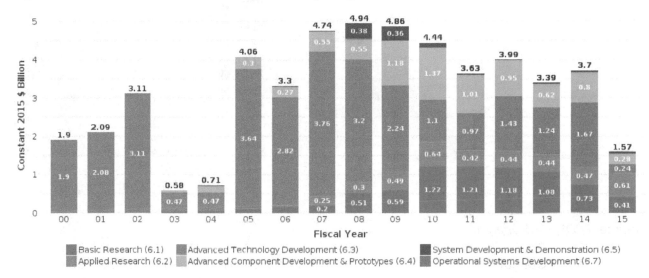

Source: FPDS; CSIS analysis.

MDA contract obligations have not followed the same pattern as the three military services during the current budget drawdown. Overall MDA contract obligations rose by more than a third between 2010 and 2013, but have fallen by 40 percent since, to their lowest levels since 2006. Meanwhile, MDA R&D contract obligations, which fluctuated around $4 billion between 2010 and 2014, plummeted by 58 percent in 2015, to the lowest level since 2004. R&D contract obligations, which had accounted for over three-fourths of overall MDA contract obligations from 2005–2010, accounted for only 34 percent in 2015, the lowest share since 2004.

The massive decline in MDA R&D is spread across MDA's R&D contracting portfolio. MDA contract obligations for Basic Research fell by 44 percent in 2015, and have fallen by nearly

---

[42] Note that, due to the massive reclassification of MDA contract data in FPDS that was discussed in Chapter 2, both the toplines and the trend lines for MDA R&D contract obligations have fundamentally changed. This change will be reflected in future CSIS analysis of MDA contract trends.

two-thirds since 2013, to their lowest level since 2007. ATD fell by an incredible 86 percent in 2015, to the lowest level in the 2000–2015 period. And ACD&P fell by 65 percent in 2015, to the lowest level since 2006. The only category of R&D with significant obligations that did not see enormous declines was Applied Research, which has risen by nearly 40 percent since 2013.

This one-year decline appears to be an artifact of the broad-based reclassification of MDA R&D contract obligations discussed in Chapter 2, rather than any actual trend; CSIS is currently engaging with DoD policy officials to better understand the scope of, and reasoning behind, this enormous change in the back-years data.

Other DoD—R&D Obligations Preserved Relative to DoD Overall

Figure 4-8: Other DoD R&D Contract Obligations by Stage of R&D, 2000–2015

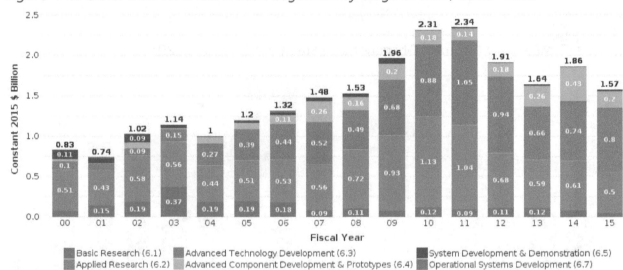

Source: FPDS; CSIS analysis.

R&D contract obligations by Other DoD contracting entities have been relatively stable during the current budget drawdown, falling by only 21 percent since 2009, roughly two-fifths the rate of decline for overall DoD R&D contracts. That decline, however, was over three times the rate of decline for overall Other DoD contract obligations between 2009 and 2015. As a share of Other DoD contract obligations, R&D has fluctuated between 5 percent and 8 percent from 2009–2015.

ATD contract obligations (17 percent) have increased significantly within Other DoD since 2009, and ACD&P (-1 percent) was relatively stable. The main source of decline within Other DoD R&D contracting was Applied Research, which fell by 47 percent between 2009 and 2015.

In 2015, Other DoD R&D contract obligations fell by 11 percent, nearly double the rate of decline for overall Other DoD contract obligations in 2015, but less steeply than overall DoD R&D contract obligations. ACD&P contract obligations plummeted by 53 percent, after tripling between 2011 and 2015, while obligations for Applied Research fell by 18 percent. Meanwhile, obligations for ATD rose by 9 percent in 2015.

As many analysts and policymakers feared, DoD R&D contracts have borne a disproportionate share of cuts within the DoD contracting portfolio during the current budget drawdown. The dimensions of those cuts, however, have not followed the expected path. Despite fears that early-stage, seed-corn R&D would be hit particularly hard, the data show that it has been relatively preserved compared to the overall declines in R&D. In fact, within DoD, two categories of mid- to late-stage R&D, Advanced Technology Development (6.3) and System Development & Demonstration (6.5) have seen cuts of two-thirds or more between 2009 and 2015.

The two main drivers of the massive declines in those two stages of R&D are the cancelation of large R&D programs (such as the Army's Future Combat Systems) and the maturation of R&D programs into procurement (such as the F-35 Joint Strike Fighter). During the budget drawdown period, however, there has been a dearth of new development programs for major weapons systems to replace those that have either graduated into production or been canceled. Even as overall R&D contract obligations have stabilized, FY2016 saw a further decline in obligations for System Development & Demonstration (6.5), indicating that the trough in DoD's development pipeline for major weapons systems has extended into its seventh year.

This trough has manifested differently within the three military services. In the Air Force, significant work and funding for the B-21 bomber is likely to begin within the next couple of years. The Navy has the Columbia-class ballistic missile submarine program on the horizon, and with its recent Milestone B certification, significant development funding should ramp up in the next few years. The Army is in the toughest position of the three, as since the failure of Future Combat Systems, the Army has been largely unable to start or sustain major development programs. With continuing uncertainty about future missions and capabilities, as well as significant budgetary challenges, the Army's trough seems likely to persist for the foreseeable future.

As part of CSIS's upcoming report on trends in federal R&D contracting, CSIS tested seven hypotheses reflecting the conventional wisdom, or at least widely expressed concerns, regarding the impact of the budget drawdown on federal R&D contracting and the supporting industrial base. For six of those seven hypotheses, the data either did not provide significant support for the hypothesis or actually strongly pointed in the opposite direction. This result underscores the importance of relying on data for analysis of trends in federal contracting; while anecdotes and the conventional wisdom may tell stories that make intuitive sense, good data is the only way to understand what is really happening. However, the overall concern that R&D contracting would be disproportionately impacted by sequestration and its aftermath was proven correct, showing the limits of management alone in mitigating the impact of the budget drawdown on U.S. technological superiority in the face of sudden, massive funding reductions.

Figure 4-9: DoD Contract Obligations by Platform Portfolio, 2000–2015

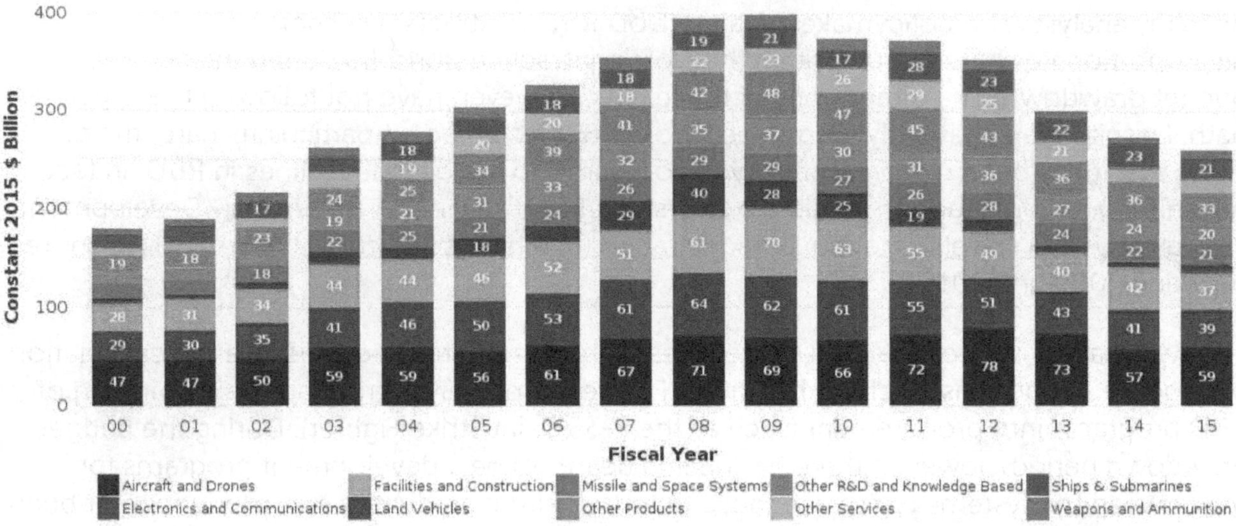

Source: FPDS; CSIS analysis.

Figure 4-9 shows Overall DoD contract obligations by platform portfolio between FY2000 and FY2015. There were no large shifts in DoD's portfolio, as most platform portfolio categories remained steady or saw relatively small (one to two percentage point) changes in the share of Overall DoD contract obligations.

Between 2014 and 2015, Land Vehicles, Weapons and Ammunition, and Aircraft and Drones all grew while Overall DoD contract obligations declined. Contract obligations for Land Vehicles grew from $5.76 billion in FY2014 to $7.30 billion in FY2015, a 27 percent growth. The growth in Land Vehicles contract is largely attributable to the $1.33 billion growth in Army Land Vehicle contract obligations as a result of increased spending on "Trucks and Tractors, Wheeled."[43] Contract obligations for Weapons and Ammunition grew 11 percent in FY2015, largely as a result of increased Air Force spending on guided missiles.[44]

Overall DoD contract obligations for Electronics and Communications (-4 percent), Missiles and Space Systems (-4 percent), and Other R&D and Knowledge Based (-7 percent) fell at rates similar to the overall rate of decline (-5 percent).

Facilities and Construction (-11 percent), Other Products (-16 percent), Other Services (-9 percent), and "Ships and Submarines" (-11 percent) declined at rates higher than the overall rate of decline. The decline in Ships and Submarines is largely the result of declines in Navy contract obligations for "Submarines" and "Combat Ships and Landing Vessels" and is explored more in-depth in Chapter 7.2. The -11 percent decline in Facilities and Construction contract obligations decline in 2015 largely resulted from a -$3.3 billion decline in Army contract obligations in that platform portfolio. Of the -$3.3 billion decline in Army Facilities and Construction contract obligations, "Construction of Other Non-Building Facilities" and "Construction of Other Administrative Facilities & Services Buildings" saw the two largest

---

[43] Product or Service Code 2320–Trucks and Tractors, Wheeled.
[44] The Air Force spent $1.08 billion in FY2015 on PSC 1410–Guided Missiles after spending only $0.02 billion in FY2014.

declines. The largest declines in Other Services contract obligations occurred in Military Health's "General Health Care Services," which fell by $0.94 billion compared to 2014. This decline returned "General Health Care Services" closer to historical levels after a one-year spike in 2014. Finally, the -16 percent decline in Other Products is the result of a $2.6 billion decline in DLA "Liquid Propellants–Petroleum Base" contract obligations compared to the previous year. As reflected in the 9 percent decline in Other Services in 2015, contract obligations for "Liquid Propellants–Petroleum Base" totaled approximately $7.5 billion as compared to approximately $11.9 billion the year before.

4.2. . Defense Contract Obligations by Budget Account

As a result of provisions in the American Recovery and Reinvestment Act of 2009 (ARRA), DoD contract data in FPDS now includes the Treasury account information needed to track which budget accounts contract obligations are funded out of. Though this capability only extends back to FY2012, the data nonetheless provides a valuable additional view into how the budgetary process influences contract obligations. Figure 4-10 shows DoD contract obligations funded out of the major DoD budget accounts between 2012 and 2015.

Figure 4-10: DoD Contract Obligations by Budget Account, 2012–2015

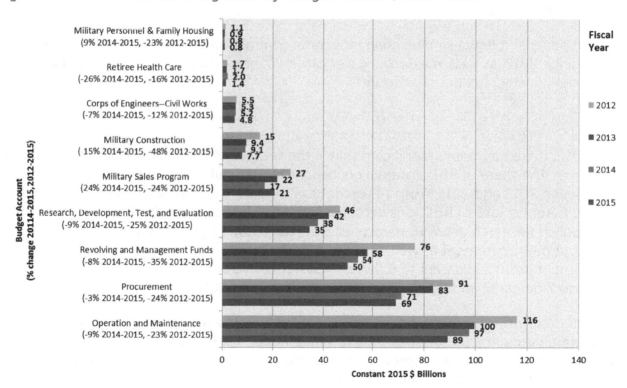

Source: FPDS; CSIS analysis.

Between 2012 and 2015, as overall DoD contract obligations declined by 26 percent, contract obligations funded out of five budget accounts saw declines roughly in line with that overall decline: Family Housing (-27 percent), the Military Sales Program (-24 percent), Operations & Maintenance (O&M) (-23 percent), Procurement (-24 percent), and Research, Development, Test, and Evaluation (RDT&E) (-25 percent). The latter four are four of the five largest budget accounts in terms of total contract obligations funded; contract obligations funded out of the fifth, Revolving and Management Funds, saw a 35 percent decline between 2012 and 2015. Contract obligations funded out of the Military Construction account (-48 percent) have declined at nearly twice the rate of overall DoD since 2012, while contract obligations funded out of Retiree Health Care (-16 percent), Military Personnel (-18 percent), and the Army Corps of Engineers — Civil Works account (-12 percent) were relatively preserved.

In 2015, as overall DoD contract obligations declined by 5 percent, obligations funded out of Military Construction (-15 percent), Retiree Health Care (-26 percent), O&M (-9 percent), RDT&E (-9 percent), and Revolving and Management Funds (-8 percent) declined more steeply than overall DoD. Contract obligations funded out of Procurement (-3 percent) and the Army Corps of Engineers–Civil Works account (-7 percent) declined roughly in line with the overall DoD rate of decline. Meanwhile, there were notable increases in obligations funded out of three budget accounts: Family Housing (6 percent), Military Personnel (12 percent), and the Military Sales Program account (24 percent).

Over the 2012 to 2015 period, the shares of contract obligations funded out of particular budget accounts did not change by more than two percentage points in any of the major DoD budget accounts.

The sections that follow examine changes in how contract obligations for R&D have been funded since 2012, as well as examining trends within the portfolios of contract obligations funded out of selected budget accounts.

4.2.1. Research & Development

Since 2012, there has been a significant shift in the sources of funding for DoD R&D contract obligations. The share of R&D contract obligations funded out of Procurement has fallen by half between 2012 and 2015, from 14 percent to 7 percent. Looking at the sources of funding for the different stages of R&D, this decline is most concentrated in mid- to late-stage R&D: while Applied Research (from 16 percent to 8 percent) saw a significant decline in the share funded out of Procurement as well, the largest declines were seen in Advanced Technology Development (from 15 percent to 5 percent), System Development & Demonstration (22 percent to 7 percent), and Operational Systems Development (13 percent to 4 percent).

This data is consistent with the findings discussed in the "Research and Development Contracting during the Budget Drawdown" section of Chapter 4 related to the now-seven-year trough in DoD's development pipeline for major weapons systems. Because a significant share of mid- to late-stage R&D tied to major platforms would be funded out of procurement, and because there has been a dearth of new major development programs over the last six years, the share of R&D funded out of Procurement has declined precipitously.

4.2.2. Procurement

Since 2012, over 80 percent of contract obligations funded out of Procurement have gone for products, and product contract obligations funded out of Procurement have declined roughly in line with the overall rate of decline for DoD products contract obligations since 2012. By contrast, services contract obligations funded out of Procurement (-13 percent) have declined notably less steeply than overall DoD services, while R&D contract obligations funded out of Procurement (-70 percent) have declined far more steeply than overall DoD R&D contract obligations since 2012.

In 2015, Procurement contract obligations for products were stable (1 percent), as with overall DoD products contract obligations. Services contract obligations funded out of Procurement (-12 percent) declined half again as steeply as overall DoD services contracts, while R&D (-46 percent) declined at nearly three times the rate of overall DoD R&D contract obligations.

4.2.3. Research, Development, Test, and Evaluation

Somewhat counterintuitively, only about half of the contract obligations funded out of the RDT&E account actually go for R&D; the remainder is roughly evenly split between products and services. Both R&D contract obligations (-30 percent) and products contract obligations (-18 percent) funded out of RDT&E were relatively preserved compared to those contract areas for DoD overall. Meanwhile, services contract obligations funded out of RDT&E (-23 percent) declined in parallel to the overall rate of decline for DoD services contracts.

In 2015, R&D contract obligations funded out of RDT&E (-15 percent) declined roughly in parallel to the overall decline in DoD R&D contracts. Products contract obligations funded out of RDT&E (-6 percent) declined moderately in a year where overall DoD products contract obligations were stable, and services contract obligations (3 percent) saw minor growth as the decline in overall DoD services accelerated in 2015.

4.2.4. Revolving and Management Funds

Approximately 70 percent of contract obligations funded out of Revolving & Management Funds have been for products in the 2012–2015 period, with most of the rest going for services. While those services contract obligations have declined (-23 percent) in parallel to overall DoD services contracts since 2012, products contract obligations funded out of Revolving & Management Funds (-40 percent) have declined over half again as steeply as overall DoD products contracts.

In 2015, services contract obligations funded out of Revolving & Management Funds (-8 percent) again declined in parallel to the overall decline in DoD services contracts, while the decline in products contract obligations (-8 percent) stands in contrast to the overall stability in DoD products contract obligations.

4.2.5. Military Sales Program

While the majority of contract obligations funded out of the Military Sales Program account unsurprisingly are for products, approximately one-quarter of contract obligations are for services. Between 2012 and 2014, services contract obligations funded out of the budget account declined by 20 percent, roughly in line with the decline in overall DoD services contract obligations over that period. In 2015, however, as the decline in overall DoD services contract obligations accelerated, services contract obligations funded out of the Military Sales Program account rose by 25 percent, due to significant increases in obligations for both Equipment-related services and Professional, Administrative, and Management Support services.

5. How Is DoD Buying It?

In this chapter, the report moves from examining what DoD is buying to how DoD buys it, which has been a subject of particular focus by the House and Senate Armed Services Committees over the past year. The first section of this chapter examines the competing reform proposals of the two congressional defense committees, and how their proposed reforms would change the structure and function of the defense acquisition system. The second section of this chapter looks specifically at issues related to contract and fee type usage in defense contracting, examining trends in how contract and fee types have been used as a way of informing the debate on how (and whether) that pattern of use needs to change. The third section of this chapter looks at policy issues specifically related to DoD's services contracting portfolio. The fourth section builds upon both past CSIS research and Under Secretary Kendall's annual Performance of the Defense Acquisition System reports, looking at what FPDS contract data can tell us about cost and schedule performance for DoD contracts. And the fifth section looks at broad trends in the level of competition for DoD contract obligations, broken down by platform portfolio.

5.1. Reforming the Defense Acquisition System

5.1.1. Acquisition Reform and Its Importance in 2016

While 2015 was a year of major activity on acquisition reform, 2016 delivered even greater changes. This is due in no small part to the desire of congressional leaders in both chambers, notably Senate Armed Services Committee Chairman John McCain and House Armed Services Committee Mac Thornberry, for a significant shake-up in the way the defense acquisition system operates. This desire is based on a belief that the system requires fundamental change, particularly in its ability to deliver timely new capabilities, and contrasts with the more incremental approach to acquisition reform of continuous process improvement that has been championed by Under Secretary of Defense for Acquisition, Technology, and Logistics Frank Kendall.

For the last several years, Secretary Carter and Under Secretary Kendall have worked within the framework of their successive Better Buying Power initiatives to incrementally improve the defense acquisition system. These initiatives came out of a recognition that DoD does not do a good enough job of managing cost, schedule, and requirements in acquisition, which has led to strained budgets and delays in getting war fighters the goods and services that they need. In a May 10, 2016, speech at CSIS on "The State of Defense Acquisition," Under Secretary Kendall stated his view that this incremental approach to improving defense acquisition was the right one and that it was bearing fruit, with data showing slowing growth in both program cost and program cycle time for Major Defense Acquisition Programs (MDAPs).[45] 2016 witnessed the implementation and maturation of the third iteration of Better Buying Power, which established a substantially sharpened focus on delivering technological

---

[45] CSIS discusses measuring the performance of the defense acquisition system in greater detail in Chapter 5.

superiority. While BBP 3.0 continued the department's incremental approach, its primary objective was substantially more ambitious than earlier versions, which focused more on improving productivity and professionalism in the defense acquisition system.

Congressional leaders signaled their willingness to break from this incremental approach in the FY2016 National Defense Authorization Act (NDAA), however, by shifting responsibility for major acquisition decisions from AT&L back to the military services, due to their belief that greater involvement by the services would increase accountability for the performance of the acquisition system. In the FY 2016 NDAA, Congress significantly expanded the responsibilities and authorities of the service chiefs and service acquisition executives (SAEs) in the acquisition process, delegating Milestone Decision Authority to the SAEs and giving the service chiefs a greater role in setting requirements and in directing tradeoffs between cost, schedule, and capability. While the increased role of the military service chiefs was implemented immediately and has already led to noticeably increased involvement by the chiefs in major acquisition issues,[46] the change in milestone decision authority was deferred until 2017. Under Secretary Kendall has expressed reservations about this proposed shift. He believes that, after the FY 2016 NDAA, the service chiefs have a sufficient role in the parts of the acquisition process where their input is most needed, such as in setting requirements, and that decentralizing the system further could undermine the recent improvement in the performance of the acquisition system.

There is great significance in the fact that an apparent fundamental divergence in approach has opened up between Congress and the former administration on acquisition reform. While acquisition reform appeared to be rare areas of opportunity for cooperation between the administration and Congress heading into 2015, it developed into an area of substantial conflict instead in 2016. And the potential effectiveness of both internal DoD and congressional reform efforts was diminished as a result. With a new presidential administration in which the president has already indicated a direct personal interest in being involved in defense acquisition, it remains to be seen whether the Trump administration will be aligned with, or will diverge from, Congress on defense acquisition.

5.1.2. Fundamental Changes in the FY2017 NDAA

Between the House and Senate versions of the FY2017 NDAA, Congress proposed fundamental changes across three major elements of the defense acquisition system: how the system is organized and given its mission; how acquisition programs are structured; and what the business model is for defense research and development. A substantial portion of these changes was included in the final version of the NDAA. However, many of the statutory changes enacted will not be fully implemented until 2018, and the desire to reshape the defense acquisition system in Congress combined with the new administration's early interest in the topic suggests the likelihood for continued debate on these fundamental issues in the coming years. The following sections outline the content and purpose of the

---

[46] Jen Judson, "US Army Chief Moves to Center of Acquisition Universe," *DefenseNews*, March 10, 2016, http://www.defensenews.com/story/defense/land/army/2016/03/10/us-army-chief-moves-center-acquisition-universe/81588944/.

three fundamental changes initially proposed in the House and Senate bills, and then describes the outcome on these issues included in the final conference agreement.

How the Defense Acquisition System Is Organized and Managed

Currently, the USD AT&L manages the defense acquisition system overseeing five assistant secretaries of defense specified in statute: acquisition; research and engineering; logistics and material readiness; nuclear, chemical, and biological defense programs; and energy, installations, and environment. The current position was established as the under secretary of defense for acquisition as part of the reforms associated with the Packard Commission and the Goldwater Nichols Act in 1987 and responsibilities for Logistics and Material Readiness were formally added in the 1990s. Prior to the creation of USD AT&L, there was an under secretary of defense for research and engineering (USD R&E), 1977–1987, which had grown out of a prior office known as the director of defense research and engineering (DDR&E). The Packard Commission recommended that the acquisition system be designed to "establish unambiguous authority for overall acquisition policy, clear accountability for acquisition execution, and plain lines of command for those with program management responsibilities" and that the system be managed by a defense acquisition executive (DAE).[47]

The Senate proposed to reestablish an USD R&E in the Office of Secretary of Defense with a central mission of leading technology innovation at DoD. To reinforce the new organization's focus on this mission, the Senate wanted to divest many of the current functions of USD AT&L to other parts of DoD by dividing current USD AT&L functions between the USD R&E and a proposed under secretary of defense for management and support (USD M&S). The Senate bill was highly specific in certain respects. For example, while today's AT&L enshrines a notional equality for the closely related functions of acquisition, research and engineering, and logistics and sustainment, the Senate bill was careful to delineate that in its proposed R&E organization, acquisition would be a subsidiary function to research and engineering, and logistics and sustainment would be a further subsidiary function to acquisition. Surprisingly, given the Senate bill's focus on elevating the research and engineering function, oversight of developmental test activities and of DoD's major range and test facilities would have been transferred from the research and engineering function and put under the control of the director of operational test and evaluation.

In many other areas, the Senate bill was nonspecific, or unclear, leaving DoD some discretion to decide how to organize. The bill divided responsibility for the logistics and sustainment function between the USD R&E, who would have managed many aspects of logistics for weapons systems, and the USD M&S, who would have had responsibility for managing the Defense Logistics Agency (DLA). DoD opposed this division of AT&L's responsibilities. Secretary Carter opposed disestablishing AT&L as an organizational step backwards and threat to recent progress in the acquisition system while conceding that organizational improvements could be made at DoD to support innovation. During a speech in May 2016, Secretary Carter stated that he too shared

---

47 David Packard, President's Blue Ribbon Commission on Defense Management, Letter to President, June 30, 1986, http://www.ndia.org/Policy/AcquisitionReformInitiative/Documents/PackardCommission-Report.pdf.

the views of the SASC that over time, the acquisition executive's position has become so preoccupied with program management, including a lot of unnecessary bureaucracy associated with it that perhaps takes some management attention away from the research and engineering function. So, I do, however, have a serious caution: separating research and engineering from manufacturing, which is implied in this proposal, could introduce problems in the transition from the research and engineering phase to the production phase and then to the sustainment phase, and that is in fact, a frequent stumbling block for programs.[48]

Changing How Acquisition Programs Are Structured

Management of defense acquisition programs has historically followed a well-defined structure. The approach is described in detail in DoD Instruction 5000.02,[49] which lays down a progression of acquisition activity from defining a need for a material solution, through technology maturation, detailed system design and development, to production and sustainment. This structure is ordered around unified acquisition programs, sometimes referred to as "programs of record." Milestones separate different phases of the program, at which point a milestone decision authority determines whether the program is ready to proceed. An acquisition program baseline is established at Milestone B in accordance with approved cost and schedule estimates. This structure provides a unifying system for coordinating acquisition activities with the requirements and budgeting processes. Requirements are established in preparation for critical milestones early in the acquisition process, and the acquisition program baseline guides the budgeting process. Much of the structure for this process is defined in statute for major defense acquisition programs (mostly included in acquisition category I in the acquisition program hierarchy), and DoD also flows down most of this structure to smaller programs (acquisition categories II and III).

The House version of the FY2017 NDAA sought to a great extent to divorce activities early in the technology development cycle from the work required to integrate an overall system design and enter production. In effect, elements of acquisition programs before Milestone B would be carried out mostly if not entirely separate from a program of record. Technologies would enter a program of record only when they have already been highly matured through a separate process. The House version sought to implement a new structure for technology development and prototyping in the form of oversight boards within each service that would guide the development of technology using funds separate from those budgeted for programs of record. This approach was intended to allow technology to develop more independently and aggressively, and feed into the existing program structure when it is ready. Because the acquisition system has been organized almost entirely around programs of record, there are currently limited resources available to develop technologies independently, and also limited mechanisms to establish requirements for the development of new technology outside the traditional programmatic structure. Some examples of this approach do exist today, however, such as advanced concept technology demonstrations.

---

[48] U.S. Department of Defense, Secretary of Defense Speech, "Remarks at Navy League Sea-AirSpace-Convention," May 17, 2016, http://www.defense.gov/News/Speeches/SpeechView/Article/774658/remarks-at-navy-league-sea-air-space-convention.

[49] U.S. Department of Defense, "DoD Instruction Number 5000.02," January 7, 2015,

The Senate bill took a different but related approach by seeking to create pathways to bypass the traditional acquisition program structure entirely. Section 804 of the FY2016 National Defense Authorization Act (NDAA) created the authority for these alternative acquisition pathways specifying a pathway for rapid prototyping and a pathway for rapid fielding. In the FY2017 NDAA, the Senate sought to further the effort by exempting programs utilizing these alternative acquisition paths from coverage under the MDAP oversight regime and by authorizing rapid acquisition funding accounts within the military services to provide resources for these nontraditional programmatic approaches. The Senate bill left unspecified how this alternative rapid-fielding system would operate differently from the traditional acquisition system. This open-ended approach would give DoD maximum flexibility in designing alternatives, but also would make it less likely that DoD will successfully overcome the inherent inertia associated with doing things outside of regular order. The Obama administration did not raise objections to either the House or Senate provisions in this area.

Changing the Business Model for Research and Development

The Senate pushed forward in the FY2017 NDAA on an effort to alter the traditional business model for defense research and development in an effort to increase industry's accountability for cost growth and to increase access, particularly for nontraditional contractors. This dual purpose was the intent behind the Senate's effort to erect new barriers to the use of cost-plus contracts and to shift responsibility for management of Cost Accounting Standards for defense contracts to a new entity within DoD.

The Senate bill proposed to put in place very strong incentives against using cost reimbursable contracts. It would have required high-level approval for the use of cost-type contracts, starting with contracts for more than $50 million and eventually covering all contracts over $5 million. If the $5 million approval threshold had been in place in 2015, it would have applied to nearly 7,400 contracts with a total value of nearly $74 billion in that year alone. The Senate bill would also have financially penalized DoD for using cost-type contracts for activities funded out of procurement and research and development accounts. The penalty would have been 2 percent of the contract amount for contracts funded by procurement, where the use of cost-type contracts is extremely rare, and 1 percent of the contract amount for contracts funded by research and development, where the use of cost-type contracts is common. In related provisions, the Senate bill also would have required the use of a fixed price development on the upcoming JSTARS (Joint Surveillance Target Attack Radar System) replacement program, and although it stopped short of requiring that the previously awarded cost-reimbursable contract for the development of the B-21 bomber be renegotiated, it would have established a unique Nunn-McCurdy process for the B-21 that requires the Air Force to manage the program like a fixed-price program. The Senate bill also required DoD to establish new cost accounting standards for cost-type contracts distinct from those currently used for all federal contracts, and required DoD to align those standards with commercial accounting standards to the maximum extent practicable. The Department of Defense objected to each of these provisions.

The Department of Defense has historically worked in partnership with its industrial base, funding research and development for defense-unique systems by reimbursing firms for their R&D expenses incurred directly on the department's behalf, and also reimbursing firms for some independently initiated R&D as an allowable overhead expense. The department has believed that firms are unlikely to invest in defense-unique systems and technologies that don't have direct commercial application without some assurance that they will achieve a return on this investment. Direct reimbursement of R&D expenses, with a provision for profit, is a straightforward way of solving this problem. Because a fully reimbursable contract is essentially a no-risk proposition for industry, the rate of profit on these contracts has historically been limited.

As DoD has sought to reach out to innovative firms in Silicon Valley and elsewhere and leverage more commercial technology in recent years, its traditional approach to R&D has appeared disconnected from the R&D business models pursued in the high-tech industry. In this market initial R&D is often funded by venture capital, and subsequent R&D is funded out of revenues, with the goal of capturing a position of advantage in global commercial markets that are exponentially larger than the DoD market. Return on investment, when such investments are successful, is orders of magnitude higher than profit levels that DoD has traditionally agreed to pay for R&D. Unsuccessful investments are terminated quickly, sometimes referred to as the "fail fast" model. These firms may be unwilling to implement the government-unique accounting systems and wait the months and years required to negotiate and sign cost reimbursable contracts, undergo government audits, and receive payment for work already performed that are required in the traditional R&D business model. While it appears likely that DoD needs to be open to different business models for R&D in order to do business with firms used to the Silicon Valley approach, it is much less clear at this point how such business models would work in the context of the defense market.

DoD has previously attempted to use fixed price contracts for the development of complex weapon systems in the 1960s and the 1980s. The history of using this approach is littered with expensive failures, as well as outright disasters, and still stands without a notable success. The failures of fixed price development of the 1980s were so painful that Congress temporarily banned the use of such contracts. The current KC-46 tanker program still has a chance to become DoD's first weapon system fixed-price development contract where massive cost increases are not paid for by DoD, but are instead born by the contractor. However, even the KC-46 example demonstrates that using a fixed price contract structure that penalizes the contractor for cost growth does not ensure that costs will stay within initial estimates.

The Senate bill operated from the premise that DoD will continue to use cost-type contracts for activities such as research and development unless presented with powerful incentives against their use. It further operated from the premise that the use of these contracts discourages participation by commercial firms and other nontraditional contractors. While both of these premises may well be correct, DoD's painful past experience with fixed price development contracts cannot easily be dismissed. There are important reasons why fixed price development contracts have led to substantial cost growth in the past, and also important reasons why DoD has ended up paying the lion's share of these cost overruns in almost every case.

5.1.3. Outcomes of 2016 Acquisition Reform

The final language of the FY2017 NDAA took action on all three of the fundamental changes put forward by the Senate and House, but as is often the case, the conference agreement is a revised and reshaped version of the original proposals. On the issue of how acquisition programs are structured, the proposals put forward by the House and Senate are adopted largely intact. This outcome reflects the significant vetting done by the House over the course of the year on these proposals, including with the administration, and the more consensus nature of the proposals themselves. On the issue of how the defense acquisition system is organized and managed, the conference agreement follows the Senate proposal to divide the functions currently included in the office of the under secretary of defense for acquisition, technology, and logistics, but in a different fashion than the Senate originally proposed. Interestingly, the conference agreement aligns the job of the newly created under secretary of defense for research and engineering (USD R&E) with the House-recommended separation of technology development projects from major defense acquisition programs. That is, the duties of the USD R&E cover the development and maturation of technology prior to the system design and development (SDD) stage, the point at which MDAPs are formally initiated. The conference agreement retains the Senate's core objective of creating a USD R&E focused on innovation, but does not put the acquisition system under that official's direct control or shift responsibilities for logistics as the Senate had proposed. Instead, the new USD R&E will focus on spurring cutting-edge technology development outside the MDAP process. The conference agreement retains a USD-level acquisition executive (the under secretary of defense for acquisition and sustainment (USD A&S)), and keeps the elements of the acquisition system unified after the beginning of SDD. It leaves most of the details of how to divide the current USD AT&L office to DoD to determine, although industrial base responsibilities are listed among the duties of the USD A&S. Both of these USDs are established as Level II executives, as is the case now for the USD AT&L, but the USD R&E is clearly indicated as the senior position and the incumbent USD AT&L can assume the USD R&E position without a separate confirmation process. These organizational changes don't take effect until February 2018, and DoD is tasked to submit an interim (March 2017) and a final (August 2017) implementation plan.

On the issue of how to change the business model for research and development, the outcome was more mixed. The conference agreement establishes a preference for fixed price type contracts and requires senior-level approval for new cost type contracts valued over $50 million starting in FY2018 (lowering to $25 million in FY2019). However, the penalties associated with the use of cost type contracts initially proposed by the Senate were dropped. The conference agreement does create a Defense Cost Accounting Standards (CAS) Board and requires both the new Defense CAS Board and the existing CAS Board to examine rewriting the CAS to more closely align with Generally Accepted Accounting Principles (GAAP) where possible. Finally, the conference agreement separates independent research and development (IRAD) costs from bid and proposal costs.

5.1.4. Outlook for Acquisition Reform in 2017

With the strong level of interest in defense acquisition manifested by President Trump during his transition, especially with respect to the replacement of Air Force One and the F-35 program, it is already clear that acquisition reform will be an area of deep interest for the new administration. While the focus of the president's interest has been most clear with respect to controlling cost growth, there are also indications that he is interested in improving capabilities in areas such as nuclear modernization and missile defense. The new administration will need to quickly grapple with the changes in the FY2017 NDAA and determine how to balance the more incremental, internal DoD approach to acquisition improvement embodied in Better Buying Power with the more fundamental shifts desired by Congress, and incorporate these approaches into its own agenda for defense acquisition.

5.2. Performance of the Defense Acquisition System

Responsible for hundreds of billions of dollars in annual spending and objectives that range from competing to outright contradictory, the performance of the defense acquisition system is challenging to summarize. CSIS gathered top experts this past May to opine on "The Big Picture of Defense Acquisition."[50] Todd Harrison looked at a key budgetary challenge that pre-dated the budget caps. Current budget projections, particularly for the Air Force, require a significant increase in defense modernization accounts to pay for current and planned major programs, even if they do not experience any overruns.[51] Pete Modigliani took a closer look at how the Defense Acquisition system could "enable innovation and rapid technology insertion." Specifically, he noted that today the main opportunities for technology insertion into systems were limited to only a fraction of the 10- to 15-year life of programs. However, in addition to suggesting a path forward to generate opportunities to deploy innovative technologies, he also cited rapid acquisition organizations across the DoD components and existing efforts to partner with industry.[52] Finally, Nancy Moore discussed the state of consolidation in the defense industrial base, with special attention to the procurement base.[53] These looks independently identify real challenges in how DoD plans and executes acquisition programs.

Narrowing the scope to whether the defense acquisition system is delivering on its promises, the performance of the Defense Acquisition system perennially attracts heated criticism and reform efforts. This past year is no exception and as Section 5.1 covers, major changes are underway. Many of the fundamental disagreements about the state of the system are irresolvable, however, because the system has multiple competing goals and different participants in the debate prioritize these goals differently. Fortunately, while they may not

---

[50] The Naval Postgraduate School 13th Annual Acquisition Research Symposium, May 4–5, 2016, https://www.researchsymposium.com/conf/app/researchsymposium/unsecured/file/6/Printer%20Friendly%20Program_13th%20Annual%20Acquisition%20Research%20Symposium.

[51] Todd Harrison, "Defense Modernization Plans through the 2020s : Addressing the Bow Wave," in *Acquisition Research Symposium* (Monterey, CA: Naval Postgraduate School, 2016).

[52] Pete Modigliani, "Speed and Agility Purpose / Outline How Defense Acquisition Can Enable," in *Acquisition Research Symposium* (Monterey, CA: Naval Postgraduate School, 2016), 2, https://www.researchsymposium.com/conf/app/researchsymposium/unsecured/file/45/Modigliani_SYM-AM-16-116.pdf.

[53] Nancy Young Moore, A. Clifford, and Judith D Mele, "Trends in the DoD Industrial Base Industry Consolidation Has Been a Trend for Many Years • Industries Tend to Consolidate over Time - Faster in," in *Acquisition Research Symposium* (Monterey, CA: Naval Postgraduate School, 2016), https://www.researchsymposium.com/conf/app/researchsymposium/unsecured/file/46/Moore_14_SYM-AM-16-117.pdf.

agree on their relative importance, critics and defenders of the system do agree on a few common metrics, namely how cost and schedule expectations compare to actual outcomes.

The Senate NDAA clearly finds the performance of the system unsatisfactory, arguing that the "U.S. military [is] falling behind technologically and that the current acquisition structure and process were significant factors in the inability to access new sources of innovation."[54] Sen. McCain detailed the metrics he used to deem the acquisition system unsatisfactory the prior November in a War on the Rocks piece. His first complaint is "[i]n constant dollars, our nation is spending almost the same amount on defense as we were 30 years ago. But for this money today, we are getting 35 percent fewer combat brigades, 53 percent fewer ships, 63 percent fewer combat air squadrons, and significantly more overhead. . . . Our declining combat capacity cannot be divorced from the problems in our defense acquisition system."[55]

However, when it comes to evaluating the defense acquisition system, overall unit cost has drawbacks as it combines a variety of factors that go well beyond the scope of the system. Most notably as Sen. McCain goes on to discuss, personnel policy, and the resultant cost per U.S. service member, is one major contributor to this change, but is beyond the scope of this report. More relevant is the tradeoff between the quantity and quality of the platform the United States acquires. Indeed, Sen. McCain favorably cites the Mine Resistant Armor Protected (MRAP) vehicle program, which emphasized quantity and speed, as "perhaps the most significant defense procurement success story of the last several decades."[56] However, he argues that the MRAP program "was produced by going around the acquisition system, not through it."[57] This point is arguable. The MRAP program was executed by long-serving government acquisition professionals using a variety of normal government contracting mechanisms and procedures, but it also utilized a variety of waivers and exceptions to normal procedures. Under Secretary of Defense Frank Kendall characterizes this approach as tailoring the acquisition system to the specific program being executed, and he incorporated the policy guidance for this kind of tailoring, along with guidance relating to several other instances of tailoring, in his update of the Department of Defense Instruction 5000.02, which governs the defense acquisition system.

However, the MRAP program is not the only indicator that the Defense Acquisition system is capable of producing larger numbers of units in time to meet demand. While the United States has chosen to focus on cutting-edge platforms for its own use, it also a major exporter. Much of that exporting is done via foreign military sales, which employ the acquisition system. As the last report on arms transfers from Stockholm International Peace Research Institute indicates, "[w]ith a 33 per cent share of total arms exports, the USA was the top arms exporter in 2011–15. Its exports of major weapons increased by 27 per cent compared with 2006–10. . . . The USA delivered major weapons to at least 96 states in 2011–15, a significantly higher number of export destinations than any other supplier."[58] Success in

[54] John McCain and Jack Reed, "National Defense Authorization Act for Fiscal Year 2017, Chairman's Summary" (Washington, DC, 2017), 5, http://www.armed-services.senate.gov/imo/media/doc/FY17 NDAA Bill Summary.pdf.
[55] John McCain, "It's Time to Upgrade the Defense Department," *War on the Rocks*, 2015, 2, http://warontherocks.com/2015/11/its-time-to-upgrade-the-defense-department/.
[56] Ibid., 3.
[57] Ibid.
[58] Aude Fleurant et al., "Trends in International Arms Transfers, 2015," February (2016): 2, https://www.sipri.org/sites/default/files/SIPRIFS1602.pdf.

arms exports does not necessarily imply that the acquisition system is effectively serving U.S. military needs, but it does undercut the idea that the system is generally incapable compared to competitors.

5.2.1. Cost Overruns and Schedules Slippages

Sen. McCain also grounds his critiques in a more concrete metric: "cost overruns and schedule delays."[59] Those two criteria certainly do not capture the entire health of the acquisition system. They cannot tell us, for example, what value the goods and services the defense acquisition system is buying provide, whether the right major defense platforms are being built, whether the defense industrial base will be robust in the future, or if innovative ideas from the commercial sector are being incorporated. Even in the information they do provide, these measures also face real challenges. There is a range of different methodologies for evaluating measures of both cost and schedule growth, and DoD's widespread reliance for these measures on nonpublic data makes it challenging to reproduce research. In response to these challenges, this section seeks to present a range of different empirically grounded perspectives on the performance of the defense acquisition system.

Apparent Cost Growth Progress Despite Unfavorable Circumstances

Improving the reliability of estimates and reducing cost overruns has been the object of many defense initiatives in the past decade, from the founding of the Office of Cost Assessment and Program Evaluation to the Better Buying Power reforms. On May 10, 2016, at CSIS, USD(AT&L) Kendall made the case that significant progress has been made. Growth in contracted cost for major programs, as measured by a 5-year moving average, dropped from over 9 percent before the BBP reforms to under 4 percent in 2015, the lowest rate in the entire 30-year period.[60] Sec. Kendall also noted an increasing percent of programs that had experienced cost reductions and a steady reduction in the number of programs costing Congressional Cost-Growth thresholds since 2009.[61]

How does this data square with Sen. McCain's critiques? Here it is helpful to turn to one of McCain's major sources, the Government Accountability Office (GAO). GAO officials presented their findings at the premier gathering of research on this topic, the Acquisition Research Symposium, hosted annually by the Naval Postgraduate School. Acquisition and Sourcing Management director Michael Sullivan noted that the Defense Acquisition system "has been on the GAO's high-risk for 24 years."[62] That history also explains the discrepancy; Sec. Kendall focuses on year-to-year changes and number of programs, while Sen. McCain emphasizes aggregate cumulative cost growth. This makes a big difference because "[o]lder programs carry a majority of the total cost and cost growth since first full estimates. Of the

---

[59] McCain, "It's Time to Upgrade the Defense Department," 3.

[60] Frank Kendall, "Better Buying Power 3.0," in *The State of Defense Acquisition* (Washington, DC, 2016), 3, doi:10.1017/CBO9781107415324.004.

[61] Ibid., 8–10.

[62] Michael J Sullivan, "U.S. Government Accountability Office Weapon Acquisition Program Outcomes and Efforts to Reform DOD's Acquisition Process Affordability Challenge," in *Acquisition Research Symposium* (Monterey, CA: Naval Postgraduate School, 2016), 5.

79 programs in the 2015 portfolio, 40 were also in the 2005 portfolio representing 80 percent of the portfolio's total acquisition cost."[63]

In year-to-year terms, when it came to cost, the GAO reported mixed but overall positive news. While slightly more programs grew in cost than fell, in dollar terms "[GAO] analysis shows that 38 programs increased their buying power in the past year and reduced procurement costs by a total of $5.4 billion. This total is the net amount of cost change given the $10.6 billion in increases due to quantity increases and the $16 billion in decreases due to other program efficiencies. . . . These buying power gains outweighed losses resulting in a net buying power gain of $10.7 billion."[64] The Better Buying Power reforms have resulted in net improvements, but these changes do not erase the cost escalation from prior decades.

Notably, this progress happened in the face of significant headwinds. David McNicol and Linda Wu found that program acquisition unit cost growth tends to be higher during "relatively constrained ones" budgetary periods as compared to "relatively accommodating." They concluded that "[t]he key point to note is that high [Program Acquisition Unit Cost] growth is not persistent, but rather episodic, and correlated with environmental factors outside of the control of the acquisition process."[65] This fatalistic take was caveated significantly at the latest acquisition research forum by David McNicol and David Tate:

> In sum, the Packard reforms of late FY 1969 worked well in essentially eliminating instances of extremely high cost growth and in that way reduced average [Average Procurement Unit Cost] growth; they were not significantly improved upon in this respect through the early 2000s; and the relaxation of [Office of the Secretary of Defense]-level oversight of the [1994 and onwards Acquisition Reform starting] years was associated with a significant number of extremely high cost growth programs and, therefore, of higher average [Average Procurement Unit Cost] growth.[66]

This method cannot be applied to the BBP period until enough time has passed that to allow many of the programs started in recent years to reach completion. However, based on the findings of the GAO and USDAT&L itself, there does appear to be progress toward the lower unit cost growth of the Packard reform and onward era. This is a pertinent finding given that the estimated cost of the acquisition portfolio already exceeds the funding available under budget caps. Thus based on the metric of cost growth, the evidence supports further incremental, rather than revolutionary, reforms.

---

[63] Ibid., 6.
[64] Ibid., 12.
[65] David L McNicol and Linda Wu, "Evidence on the Effect of DoD Acquisition Policy and Process on Cost Growth of Major Defense Acquisition Programs" (Alexandria, VA, 2014), 14, http://www.acq.osd.mil/parca/docs/ida-p5126.pdf.
[66] David McNicol and David Tate, "Further Evidence on the Effect of Acquisition Policy and Process on Cost Growth," in *Acquisition Research Symposium* (Monterey, CA, 2016), 145.

While both the House and Senate acquisition reform proposals call for programs that are delivered "on time and on-cost,"[67] Rep. Thornberry's bill was particularly concerned with schedule. The Chairman's Memo repeats a concern from 2016 that "the conventional acquisition system of the [DoD] is not sufficiently agile to support warfighter demands."[68] The fear is not surprising, as the GAO reported that in 2015 the "average delay in delivering initial capabilities has increased to almost 30 months."[69] However, as the Cover Memo for the House acquisition proposal points out, schedule growth is only part of the problem: "On average, major defense acquisition programs operate for 9 years before yielding new capabilities. Requirements determination, budgeting, and contracting can each take another 2 years or more before programs begin."[70]

Digging deeper into the GAO's bad news, the 29.5-month average delay reported by GAO for 2015 represented an additional 2.4-month delay versus 2014 and "continues a trend we have seen for the past decade."[71] Does this mean that the acquisition system is just taking longer to deliver than in the recent past? The answer, surprisingly, is no. David Tate measured cycle time, the number of years from program initiation to the Initial Operating Capability (IOC), and compared how today's program compared to those of the past. "Going back to the late 1980s, there is no apparent upward trend [in cycle time]. Statistical analysis confirms that the trend is indistinguishable from zero, and that the median cycle time has been roughly eight years over that entire span."[72] He does note one significant caveat, "there is a noticeable upward trend for the programs that are spending the most money on procurement."[73] Nonetheless, the absence of a slowdown is echoed by Jennifer Manring and Thomas Fugate, who studied the briefer period between Milestone B to Milestone C. They found that the gap between the two milestones was shorter than in prior years, attributing it to the fact that "large acquisition programs have trended away from single pass (aka, "Big Bang") efforts in favor of incremental development and delivery of needed capabilities."[74] Looking at contract cycle times rather than program cycle times, Sec. Kendall similarly found that development and pre-procurement contracts grew significantly shorter in duration over the past decade plus.[75]

---

[67] Mac Thornberry, "Thornberry - Acquisition Agility Act Cover Memo.pdf" (Washington, DC, 2016), 1, https://armedservices.house.gov/sites/republicans.armedservices.house.gov/files/wysiwyg_uploaded/Acquisition Agility Act Cover Memo.pdf. The Senate summary uses similar language: "on time, on budget, and as promised." McCain

[68] Thornberry, "Thornberry - Acquisition Agility Act Cover Memo.pdf," 3.

[69] Sullivan, "U.S. Government Accountability Office Weapon Acquisition Program Outcomes and Efforts to Reform DOD's Acquisition Process Affordability Challenge," 6.

[70] Thornberry, "Thornberry - Acquisition Agility Act Cover Memo.pdf," 3.

[71] Sullivan, "U.S. Government Accountability Office Weapon Acquisition Program Outcomes and Efforts to Reform DOD's Acquisition Process Affordability Challenge," 10.

[72] David Tate, "Acquisition Cycle Time: Defining the Problem," in *Acquisition Research Symposium* (Monterey, CA: Naval Postgraduate School, 2016), 74.

[73] Ibid., 75.

[74] Jennifer E. Manring and Thomas M. Fugate, "Schedule Analytics," 2016, 91, https://www.researchsymposium.com/conf/app/researchsymposium/unsecured/file/117/Manring,Fugate_SYM-AM-16-028.pdf.

[75] Kendall, "Better Buying Power 3.0," 11.

This contrasting data raises the question why is there so much schedule growth if cycle times are not increasing? Given the longstanding acquisition bow wave, it is not surprising that some programs have experienced schedule slippages to keep annual costs down, particularly since budget caps were put in place. However, Tate argues the larger problem is that schedule estimates are not calculated with the same rigor as cost estimates:

> Not infrequently, the initial schedule *estimate* for an MDAP is not an estimate at all, but a constraint set externally with little regard to program content or historical precedent. Sometimes this is driven by anticipated external demands for a system that is to be used on multiple platforms, as was the case for several of the Joint Tactical Radio System (JTRS) subprograms. Sometimes it is driven by a planned retirement agenda for existing systems, such as the plan for the Global Hawk Block 30 aircraft to replace the U-2. Sometimes it seems to be driven by impatience; the Army's never-quite-started Ground Combat Vehicle program was told the delivery date of the first production vehicle in its Initial Capabilities Document before even a design concept had been identified.[76]

Present goals for acquisition reform go beyond fixing schedule estimates to qualitatively changing how DoD structures programs in order to develop and update them in smaller chunks. There are a range of approaches to this goal, including Rep. Thornberry's proposal, the suggestions by Pete Modigliani, and Andrew Hunter's adaptive-systems. These proposals would not be without recent precedent. The MRAP acquisition program quickly adapted a variety of off-the-shelf platforms and designs to the battlefield to mitigate the threat of IED attacks. Hunter also points to the example of "the Predator drone, which has been continuously modified, upgraded, and morphed into new variants to respond to new threats and new technology."[77]

Despite these past successes, the schedule slippages discussed above mean that these proposals will need to bear the burden of both improving estimation on top of their other goals. Appropriately, each of the proposals discusses the handling of estimates and baselines. Rep. Thornberry's Section III covers authorities and coordination, Hunter's emphasizes dynamic but accountable baselines, and Modigliani suggests that goals for initial operating capability should be established early and prominently. Reformers can take some comfort in research that shows better estimates are possible. David Tate identifies as a starting point that "[c]ycle time growth has been increasing, especially in C3I and Space programs. Much of this growth seems to be associated with overly optimistic schedule estimates."[78] The findings regarding schedule do not point to a single solution, but does document the alarm bell to which reformers are responding.

---

[76] Tate, "Acquisition Cycle Time: Defining the Problem," 78–79.

[77] Andrew P. Hunter, "The US Needs More Weapons That Can Be Quickly and Easily Modified," *Defense One*, April 2016, 1, http://www.defenseone.com/ideas/2016/04/us-needs-more-weapons-can-be-quickly-and-easily-modified/127787/.

[78] Tate, "Acquisition Cycle Time: Defining the Problem," 87. Optimistic scheduling is earlier defined as those programs expected to take fewer years than the average number of years for that sort of platform.

5.2.2. Metrics from the Contract Data

Up to this point, this section's evaluation of the performance of the Defense Acquisition system focused on major weapon systems. Those programs are responsible for a significant percentage of DoD contract spending, but are only a subset of the defense contracting system. DoD also obligates billions for smaller programs and even more for operations and maintenance work.

To identify techniques to better measure, and thus understand and manage, the defense contracting system, last year CSIS published "Avoiding Terminations, Single-Offer Competition, and Costly Changes with Fixed-Price Contracts," a report that featured two new approaches to looking at contract performance: contract failure as measure by termination rates and cost-growth as measured by contract ceiling breaches. The study found that acquisition officials were already effectively balancing contract choices type in areas where fixed-price contracts where higher risk, with the exception of long duration contracts. Nonetheless, fixed-price contracts were still across the board twice as likely to be terminated as cost-based contracting, suggesting that the contract pricing method can control cost but faces greater challenges adapting when difficulties arise. This year, thanks to the support of the Naval Postgraduate School, CSIS extends this approach to analyzing trends within the entire defense contracting system.[79] This preliminary look still faces limits in the quality of the underlying data, but past experience has shown that the best way to improve data quality is to make transparent use of it. In the spirit of the NPS sponsorship that initially created this research, CSIS also shares the dataset online for other researchers interested in conducting their own analysis.[80]

Partial and Complete Terminations

If a partial or complete termination has happened in an acquisition program, something has gone wrong. There's a wide range of potential causes: the contractor may not have performed adequately, the needs of the customer may have changed dramatically, or a bid protest may have overturned a contract award. The information provided is limited, for example FPDS does not differentiate between when part or when all of a contract is terminated. Despite these limitations, even a partial termination is enough typically of bureaucratic hurdle and disappointment to the contractor to represent a real failure.

When looking at terminations to measure the performance of the defense acquisition system, this report is more interested in the year a contract was started than the year a termination occurred. It is true that terminations may occur due to events beyond the scope of the acquisition system to anticipate or by new mistakes in old contracts. However, many acquisition best practices emphasize improving key decisions made before entering a binding contract. For this reason, Figure 5-1 tracks the annual number of terminations, based on the start year of the contract. Because contracts with smaller ceilings account for the

---

[79] This analysis was conducted in partial support of a Naval Postgraduate School grant regarding crisis-funded contracts. That study is comparing the performance of those contracts in both the civilian and defense sphere, with more traditional contracting activity and is currently ongoing.

[80] The study team has published the dataset in greater quantitative analysis detail online at https://github.com/CSISdefense/Crisis-Funding/. For assistance in using the dataset, please contact Gregory Sanders at GSanders@csis.org.

overwhelming majority of contracts numerically, but a minority of all contract obligations measured by dollar value, the three columns in the graph segment data by contract ceiling. Similarly, longer-duration contracts are less common but are difficult to evaluate for recent years, so the three rows in the graph segments the data by initial contract duration. The total count of contracts and sum of obligations are labeled in each of the graph facets, to put the comparative importance of each category in context.

Figure 5-1: Number of Terminations for Contracts and Task Orders Starting in 2007–2014

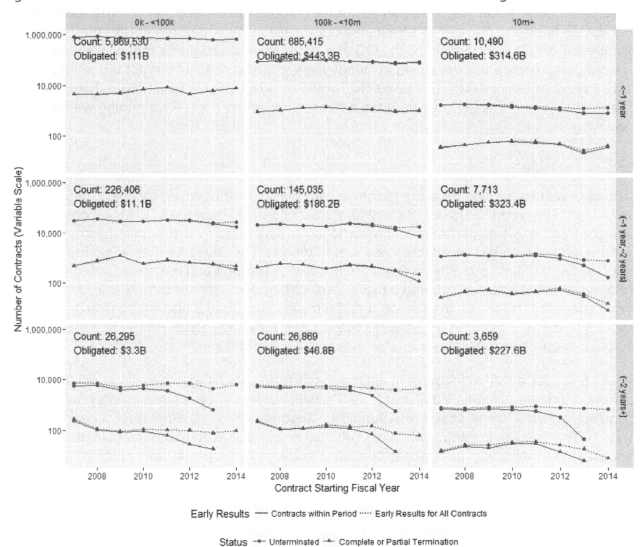

Source: FPDS; CSIS analysis.

For those contracts with initial durations of less than two years, the number of terminations typically peaks in the period between 2009 and 2012. Thus the observations for contracts and task orders has some similarities, which is partially in keeping with the trend for the cancelation of major weapon projects observed by David McNicol: "Annual average cancellation rates are much higher when procurement funding is sharply decreasing than when it is stable or increasing."[81] That said, the number of terminations is fairly volatile, often exhibiting more than one peak and trough in a span of less than a decade.

The second trend is still preliminary, the dramatic drop-off in numbers of terminations for contracts and task orders with initial durations of at least year and ceilings of at least $100K. This observation is only preliminary because many of these contracts are still ongoing. The solid blue and red lines include only contracts that have reached both their initial and their final completion date. By contrast, the dotted line includes all contracts and task orders that started in a given fiscal year, whether or not they would have ended before 2016. Looking at the dotted lines, the number of terminated contracts has been declining faster than the number of overall contracts in both 2013 and 2014. This data suggests that acquisition system performance has improved in recent years as measured by contract terminations. If future years of data validate this trend, this would be a remarkable result given that the change happened in a period where the defense budget and contract spending were both declining.

Ceiling Breaches

Ceiling breaches for contracts are roughly analogous to cost overruns for major defense acquisition programs. CSIS defines a ceiling breach to be an increase in the total cost ceiling of a contract, made as part of a change order. This is a cruder measure than the cost overruns measured by the selected acquisition reports. The cause for the ceiling breach is not clearly classified, quantity changes are not always relevant and are not adjusted for, inflation is not accounted for, and a higher cost ceiling does not necessarily translate into more spending. However, the reason that the measure is crude is also the reason it is important that contracts as a whole are subject to significantly less scrutiny than major defense acquisition programs.

Figure 5-2 uses a risk-management approach to looking at changes in ceiling breaches for contracts started between 2007 and 2014. Perhaps surprising cynics, the median defense contract, across a range of size and duration categories, does not experience a ceiling breach. This median contract, which can also be referred to as the 50th quantile, thus does not assist in monitoring for changes over time, because it remains at zero percent. However, it is easy to imagine a more cautious approach. Instead of asking whether half of all contracts avoided ceiling breaches, the study team also looked at the 80th, 90th, and 95th quantiles. This means determining the level of ceiling breaches that four out five, nine out of ten, and nineteen out of twenty contracts stayed below, respectively.[82] Strikingly, the results further

---

[81] David L McNicol, "Further Evidence on the Effect of Acquisition Policy and Process on Cost Growth, Presentation," in *Acquisition Research Symposium* (Monterey, CA: Naval Postgraduate School, 2016), 18.
[82] This approach has two advantages over the alternative of determining the mean ceiling breach. The first is that the mean value is dominated by a small number of outlier contracts that experienced exponential cost increases. These changes may simply reflect the fact that cost ceiling data is not as closely tracked as actual obligation data, meaning that a small number of extreme contracts may simply reflect errors rather than genuinely significant contracts. Secondly, the quantile approach also adjusted for the fact that a large portion of cost-ceiling changes actually reduce the cost ceiling. Thus the 1st quantile ceiling breach in many categories is actually negative. These changes may reflect terminations, accounting fixes, or the decision to wind down a contract earlier. Thus these reductions are often not comparable to achieving a cost savings, and it does not necessarily make sense to treat ceiling reductions any differently than contracts that never experienced a change order at all.

support the premise that the performance of the defense acquisition system has improved its ability to estimate costs.

Figure 5-2: Ceiling Breach by Quantile for Contracts and Task Orders Starting 2007–2014

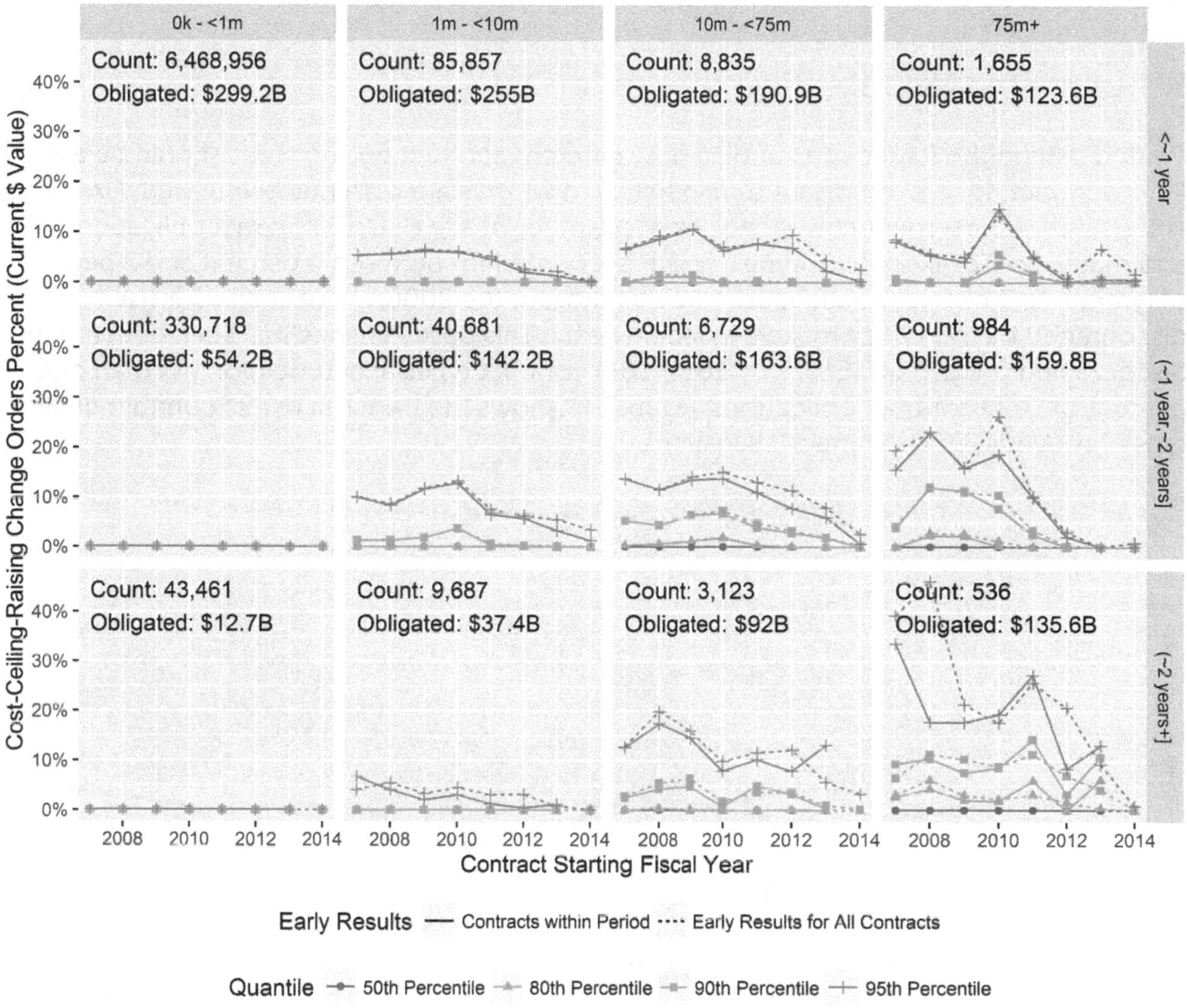

Source: CSIS; FPDS analysis.

Contracts and task orders with a ceiling of less than $1 million avoided ceiling breaches more than 95 percent of the time throughout the entire period. For contracts with higher ceilings, ceiling breaches are more prevalent. However, the good news is that for eight of the nine categories, ceiling breaches peaked somewhere between 2009 and 2010. The ninth category, contracts with ceilings of $75 million or more and a planned duration of two years, is naturally the most volatile because it has the fewest contracts. In fact, that category has less than 10 contracts that started in FY2013 that took place entirely within the study period. While that small number of contracts does show backsliding, the early results for all contracts started during that year does show a downward trend.

This finding does not guarantee that DoD is getting a good value for its money nor does it tell us much about that 5 percent of contracts in each category experiencing the largest breaches. Nonetheless, these results lend credence to the idea that cost estimating is improving and that the improvement is not just limited to those large programs that receive the most scrutiny.

5.3. Contract and Fee Type

5.3.1. Use of Contract Pricing Types in DoD

To understand the issues surrounding how different contract pricing types should be used in DoD contracting, it is important to understand how they are currently being used. Fixed price and cost reimbursement contract types account for nearly all DoD contract obligations; time and materials (T&M) contract types, which accounted for between 4 percent and 5 percent of overall DoD contract obligations through most of the 2000s, have seen their used discouraged by top policy makers in recent years. This policy change has been effective in discouraging the use of T&M — by 2014, T&M contract types accounted for less than one percent of DoD contract obligations. Figure 5-3 shows the trends in use of contract pricing types for overall DoD contract obligations.

Figure 5-3: DoD Contract Obligations by Contract Pricing Type, 2000–2015

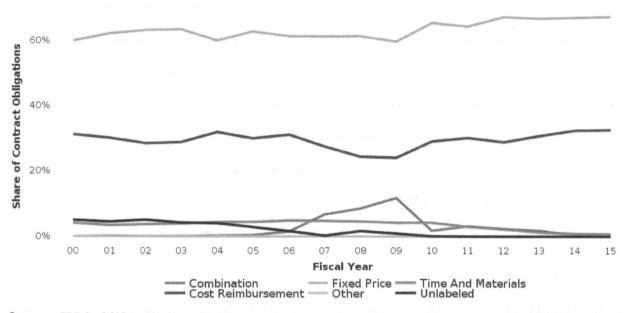

Source: FPDS; CSIS analysis.

The data show that, excluding the period where combination was masking a roughly equal mix of fixed price and cost reimbursement contracts, use of the two major contract pricing types has been remarkably consistent since 2000. Roughly 30 percent of DoD contract obligations were structured as cost reimbursement contract types in most years, while between 60 percent and 67 percent of obligations were structured as fixed price contract types. This stability is notable, given that there have been statutory, regulatory, and policy preferences in favor of fixed price contracting in place for the last several years.

While the overall usage of fixed price and cost reimbursement contract types has been highly stable, there have been significant shifts in the patterns of use for different fee types for fixed price and cost reimbursement contracts during the 2000–2015 period. For fixed price contract types, Firm Fixed Price has accounted for over 80 percent in all but two years during the 2000–2015 period. Despite that consistency, there have been notable shifts in the use of different fixed price fee types during the period. Fixed Price with Economic Price Adjustment, which accounted for between 8 percent and 12 percent of fixed price contract obligations from 2000–2013, has fallen to 5 percent by 2015. Meanwhile, Fixed Price Incentive Fee, which fell from 8 percent of overall fixed price contract obligations in 2000 to 2 percent in 2008, has risen steadily since, to 14 percent in 2014 and 2015; this is in line with the findings from Under Secretary Kendall's 2014 Performance of the Defense Acquisition System report recommending increased usage (when appropriate) of incentive-type contracts, because they motivate vendors to seek cost reduction better than other fee types.

Interestingly, there has not been a corresponding increase in the share of cost reimbursement contract obligations structured as Cost Plus Incentive; in fact, that share has declined in recent years, from 23 percent in 2011 to 15 percent in 2015. Cost Plus Award Fee, which accounted for between 40 percent and 49 percent of all cost reimbursement contract obligations between 2000 and 2007, has declined steadily since, falling to 11 percent of overall cost reimbursement contract obligations by 2015. In its place, use of Cost Plus Fixed Fee has risen from 36 percent of cost reimbursement contract obligations in 2007 to 67 percent in 2015.

By Component—Broad Stability in Use of Contract Pricing Types within Major DoD Components

The consistency in the rates of use for fixed price and cost reimbursement contracts types within DoD overall largely holds true for the major DoD components, as seen in Figure 5-4.

Figure 5-4:Contract Pricing Type for DoD Contract Obligations by Component, 2000–2015

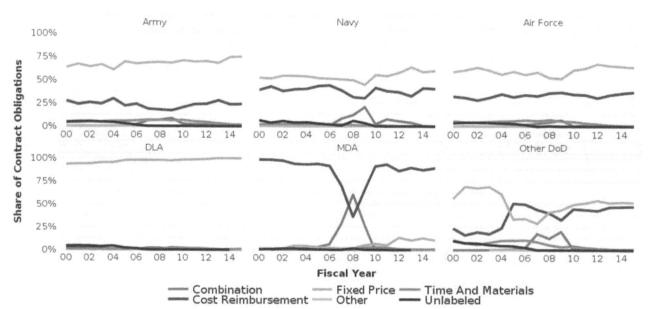

Source: FPDS; CSIS analysis.

For the Army, approximately two-thirds of contract obligations have been structured as fixed price contract types in most years during the 2000–2015 period, with approximately a quarter structured as cost reimbursement contract types. Similarly, within the Air Force, approximately 60 percent of contract obligations have been structured as fixed price contract types in most years, while roughly one-third have been cost reimbursement.

The split is more even within the Navy's contracting portfolio; roughly half of Navy contract obligations from 2000–2011 were structured as fixed price contract types, while the share structured as cost reimbursement contract types hovered near 40 percent over that period. Over the last four years, however, the share structured as fixed price contract types has risen to around 60 percent, while the share going to cost reimbursement contract types has fallen to roughly one-third.

DLA and MDA have seen notably different trends, due to the particular nature of what those two components contract for. Virtually all DLA contract obligations have been structured as fixed price contract types throughout the 2000–2015 period, while 90 percent or more of MDA's contract obligations have been structured as cost reimbursement contract types throughout the period, excluding the years when the combination classification was in use.

By Area

A major determinant of what contract pricing mechanism is used in DoD contracting is what is being contracted for, as seen in Figure 5-5.

Figure 5-5: Contract Pricing Type for DoD Contract Obligations by Area, 2000–2015

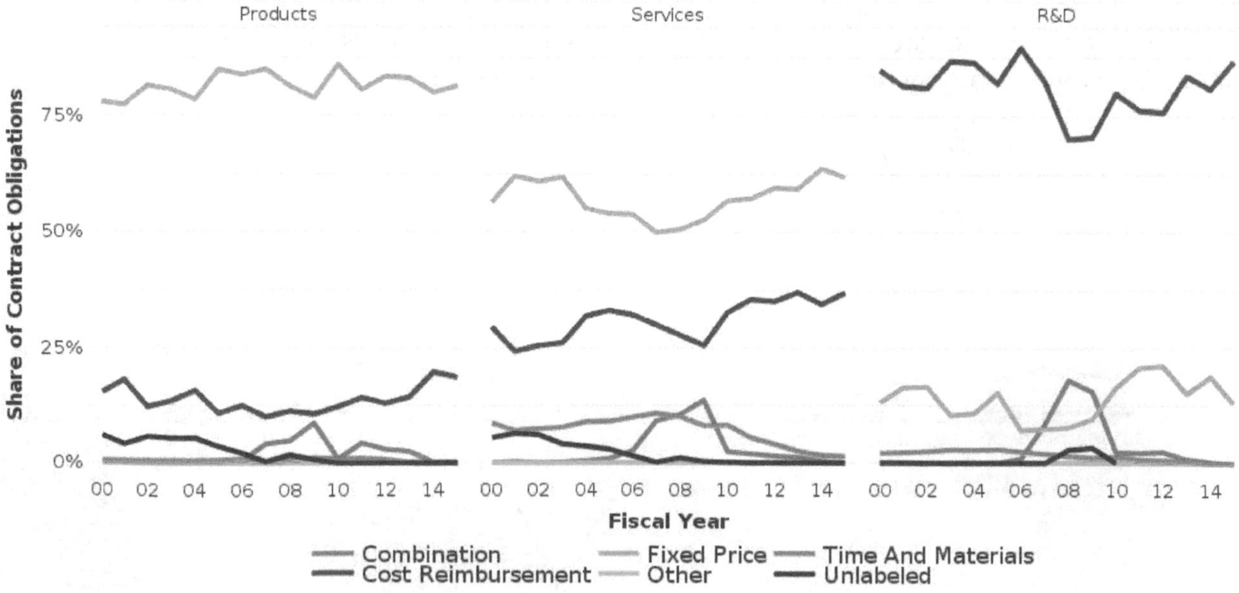

Source: FPDS; CSIS analysis.

Products—Use of Firm Fixed Price and Cost Plus Award Fee Contract Types Decline Significantly

In most years since 2000, 80 percent or greater of DoD products contract obligations have been structured as fixed price contract types, with the vast majority of those being Firm Fixed Price. Among the different categories of products, Ships and Missiles & Space were the only two that saw notably lower usage of fixed price contract types; for Ships, a majority of contract obligations were actually structured as cost reimbursement contract types in 2014 and 2015.

The share of fixed price contracts for DoD products structured as Firm Fixed Price has dropped significantly since 2012, from 81 percent to 63 percent in 2015, while Fixed Price Incentive has grown from 7 percent in 2012 to 24 percent in 2015. There has been a similar shift within the 10–15 percent of DoD products contract obligations structured as cost reimbursement contract types: the share of those cost reimbursement contracts structured as Cost Plus Award Fee fell from 41 percent in 2007 to 4 percent in 2015, while use of Cost Plus Fixed Fee (35 percent to 64 percent) has risen dramatically. Interestingly, use of Cost Plus Incentive Fee rose dramatically in the late 2000s, to a high of 43 percent in 2011, but has fallen back since, to 25 percent in 2014 and 2015.

Services—Massive Shift from Cost Plus Award Free to Cost Plus Fixed Fee

Usage of contract pricing mechanisms is more evenly distributed for DoD services contracts than for products, but the majority of contract obligations for services are still structured as fixed price contracts in every year from 2000–2015. Cost reimbursement has accounted for between a quarter and a third of DoD services contract obligations in most years during the period, reaching a peak of 37 percent in 2013 and 2015. T&M, which peaked at 11 percent of DoD services contract obligations in 2007, accounted for just over 1 percent in 2015.

Within the five categories of services, both Facilities-related Services & Construction (FRS&C) and Medical services deviated wildly from the trend for overall DoD services: over 80 percent of FRS&C contract obligations were structured as fixed price throughout the period; for Medical services contracts, over 95 percent were structured as fixed price from 2000–2003, but by 2005, 75 percent were structured as cost reimbursement, and cost reimbursement contract types have predominated since. Meanwhile, for Professional, Administrative, and Management Support (PAMS) services contracts, there has been a roughly even split between fixed price and cost reimbursement in most years, especially in recent years since use of T&M has declined.

When looking at fixed price fee types used in DoD services contracting, there has been little change over the 2000–2015 period; since 2003, over 90 percent of fixed price contracts for DoD services have been structured as Firm Fixed Price in every year. By contrast, there have been significant shifts in the types of cost reimbursement contracts used for DoD services: since 2007, the share structured as Cost Plus Award Fee has fallen from 40 percent to 14 percent, while the share structured as Cost Plus Fixed Fee has risen from 31 percent to 71 percent. Similar to what was seen for DoD overall, there has actually been a decline in the use of Cost Plus Incentive in recent years, from 23 percent in 2010 to 12 percent in 2015.

Unsurprisingly, the vast majority (over 80 percent) of DoD R&D contract obligations have been structured as cost reimbursement contract types in the 2000–2015 period. There was a brief period in the early 2010s when the use of fixed price contract types for R&D rose dramatically (from 10 percent in 2009 to 21 percent in 2011 and 2012), in response to policy guidance to increase use of fixed price contract types in Better Buying Power 1.0, but that use has fallen off since, to 13 percent in 2015. This is true across all stages of R&D, albeit with some significant year-to-year variability.

Looking at the types of cost reimbursement contract structures used for DoD R&D contracts, there has been a significant decline in the use of Cost Plus Award Fee, which has fallen from 51 percent in 2006 to just 12 percent in 2015. Three other cost reimbursement fee types saw notable increases in usage over that period: Cost No Fee (3 percent to 12 percent); Cost Plus Fixed Fee (42 percent to 64 percent); and Cost Plus Incentive (4 percent to 11 percent, though down from a high of 13 percent in 2013). For DoD R&D contracts structured as fixed price, around 90 percent were Firm Fixed Price from 2000–2013, but in 2014, 44 percent were structured as Fixed Price Incentive. That share dropped to 25 percent in 2015, but that still represents a fivefold increase over the share in 2013, indicating that the policy preference for Fixed Price Incentive has been successfully implemented.

5.3.2. Contract Pricing Mechanism by Budget Account

Because the push to increase the use of fixed price contracting is coming from Congress, and Congress's view of defense contracting necessarily runs through the prism of the budget process, it is useful to look at how contract pricing mechanism use differs for contracts funded out of different budget accounts. As mentioned in the section on overall contracting by budget account in Chapter 4, the data necessary to cross-walk contract data and budget data is only available and reliable starting in 2012, so this analysis will focus on the 2012–2015 period.

Figure 5-6 shows contract pricing mechanism use for contracts funded out of the five budget accounts that account for the largest shares of DoD contract obligations: Military Sales Program O&M, Procurement, RDT&E, and Revolving & Management Funds.

Figure 5-6: Contract Pricing Mechanism by Budget Account, 2012–2015

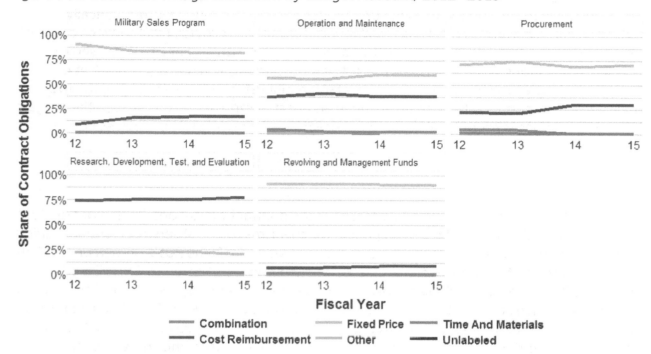

Source: FPDS; CSIS analysis.

Research, Development, Test & Evaluation—Mix of Contract Types Differs Significantly from DoD Overall for Contracts Funded out of RDT&E

As mentioned in the section on overall contracting by budget account in Chapter 3, R&D contracts have accounted for only 52 percent of the contract obligations funded out of the RDT&E account since 2012. 88 percent of those R&D contract obligations were structured as cost reimbursement, with the majority (53 percent) as Cost Plus Fixed Fee. Cost Plus Award Fee accounts for 23 percent, while both Cost No Fee (10 percent) and Cost Plus Incentive (13 percent) accounted for significant shares. For the 11 percent of RDT&E R&D contract obligations structured as fixed price, 81 percent were structured as Firm Fixed Price, while 17 percent were structured as Fixed Price Incentive. Interestingly, the share of fixed price RDT&E R&D contract obligations structured as Fixed Price Incentive rose from 8 percent in 2013 to 27 percent between 2014 and 2015, indicating that the push for greater use of incentive fee types may be having an effect.

Products have accounted for 26 percent of contract obligations funded out of RDT&E since 2012. In stark contrast to overall DoD products, which are overwhelmingly structured as fixed price, 61 percent of products funded out of RDT&E are structured as cost reimbursement, compared to only 37 percent for fixed price. Cost Plus Incentive (46 percent) and Cost Plus Fixed Fee (39 percent) are the predominant cost reimbursement fee types for RDT&E products; use of Cost Plus Fixed Fee has increased notably between 2012 and 2015, from 33 percent to 47 percent. For fixed price RDT&E products contract obligations, there is an almost even split between Firm Fixed Price (48 percent) and Fixed Price Incentive (52 percent); the latter is over two-and-a-half-times higher than the usage rate of that fee type in DoD R&D contracts overall.

Services have accounted for 22 percent of contract obligations funded out of RDT&E since 2012, and as with products, the usage of contract pricing mechanisms is dramatically different than for overall DoD services. Whereas 61 percent of overall DoD services contract obligations since 2012 have been structured as fixed price, 61 percent of RDT&E services contract obligations were structured as cost reimbursement. Additionally, 9 percent of RDT&E services were structured as T&M, by far the highest for services in any of the five largest budget accounts. For both fixed price and cost reimbursement, however, the usage of different fee types is roughly in line with the usage seen for overall DoD services contracts.

Operations & Maintenance—Use of Cost Plus Fixed Fee Has Grown to Dominate O&M Services Contracts since 2012

Services have accounted for over 80 percent of the contract obligations funded out of O&M since 2012, and the usage of contract pricing mechanisms for O&M services contracts differs slightly from the usage for DoD services overall: between 2012 and 2015, 56 percent of O&M services contract obligations were structured as fixed price, compared to 61 percent for overall DoD services over that same period. As with overall services, over 90 percent of fixed price O&M services contract obligations have been structured as Firm Fixed Price. For the 40 percent of O&M services structured as cost reimbursement, there have been notable shifts between 2012 and 2015 in the cost reimbursement fee types used: the share of cost reimbursement O&M services contract obligations structured as Cost Plus Award Fee has fallen from 23 percent to 15 percent, and the share structured as Cost Plus Incentive has fallen from 21 percent to 9 percent. Those cost reimbursement fee types have been replaced by Cost Plus Fixed Fee, which has risen from 49 percent to 73 percent.

Products accounted for 15 percent of the contract obligations funded out of O&M since 2012, and the usage of contract pricing mechanisms is roughly in line with the usage in overall DoD products contracting: since 2012, 79 percent of O&M products contract obligations were structured as fixed price, compared to 82 percent for DoD products overall. 93 percent of fixed price O&M products contract obligations were structured as Firm Fixed Price, compared to 69 percent for DoD products overall; Fixed Price Incentive, which has accounted for 20 percent of overall fixed products contract obligations since 2012, has accounted for only 5 percent of fixed price O&M products contracts.

R&D has accounted for 4 percent of contract obligations funded out of O&M since 2012. The usage of contract pricing mechanisms in O&M R&D contract obligations differs slightly from overall DoD R&D: since 2012, 23 percent of O&M R&D contract obligations have been structured as fixed price, compared to 17 percent for DoD R&D overall. For fixed price O&M R&D contract obligations, 83 percent were structured as Firm Fixed Price, with the next-highest proportion being 11 percent for Fixed Price Level of Effort. This differs notably from fixed price R&D contracts in DoD overall: only 76 percent are structured as Firm Fixed Price, while Fixed Price Incentive accounts for 21 percent. For cost reimbursement O&M R&D contract obligations, the differences are even more stark: 74 percent of cost reimbursement O&M contract obligations have been structured as Cost Plus Fixed Fee since 2012, compared to only 60 percent within overall DoD R&D. Cost No Fee, Cost Plus Award Fee, and Cost Plus Incentive all accounted for over 10 percent of overall DoD R&D contracts structured as cost reimbursement, whereas for O&M R&D, only Cost Plus Award Fee exceeded 10 percent for the 2012–2015 period.

Procurement—In Contrast to Overall DoD Services Contracts, Cost Reimbursement Is the Norm for Services Contracts Funded out of Procurement

Since 2012, 83 percent of contract obligations funded out of the Procurement account have been for products. 78 percent of those Procurement products contract obligations have been structured as fixed price over the 2012–2015 period; 60 percent of those fixed price Procurement products contract obligations were structured as Firm Fixed Price, compared to 69 percent for overall DoD products. Nearly all of the remaining fixed price Procurement products contracts were structured as Fixed Price Incentive (38 percent), which is almost twice as high as the share for overall DoD products. For cost reimbursement Procurement products contracts, there has been a significant shift in the cost reimbursement fee types used since 2012: the share structured as Cost Plus Fixed Fee has risen from 58 percent to 73 percent, while the share structured as Cost Plus Incentive has fallen by half (34 percent to 17 percent.)

For services, which have accounted for 13 percent of contract obligations funded out of the Procurement account since 2012, the split of fixed price and cost reimbursement is nearly reversed from overall DoD services: in Procurement, 55 percent of services contract obligations have been structured as cost reimbursement and 38 percent fixed price, while for DoD services overall, those shares are 36 percent and 61 percent, respectively. As with overall DoD services, nearly all fixed price Procurement services have been structured as Firm Fixed Price since 2012; the usage of different types of cost reimbursement fee types is also roughly in line with overall DoD services. There is, however, a notable difference in trends for usage of cost reimbursement fee types: while overall DoD services have seen a near-doubling of the use of Cost Plus Fixed Fee, and concurrent decline in the use of Cost Plus Incentive, Procurement services have seen a small decline in the use of Cost Plus Fixed Fee, and a tripling in the use of Cost Plus Incentive.

R&D accounts for only 4 percent of the contract obligations funded out of the Procurement account, but the split of fixed price and cost reimbursement is highly unusual: for the 2012–2015 period, Procurement R&D contract obligations were nearly evenly split between cost reimbursement (51 percent) and fixed price (46 percent), though those rates fluctuated wildly from year to year.

Revolving & Management Funds—Use of Fixed Price with Economic Price Adjustment Plummets Since 2012

Since 2012, 70 percent of the contract obligations funded out of the Revolving & Management Funds account have been for products, and those have almost entirely been structured as fixed price. 67 percent of those fixed price contract obligations have been structured as Firm Fixed Price, while 32 percent have been structured as Fixed Price with Economic Price Adjustment. There has been a significant shift in those shares since 2012, however: use of Firm Fixed Price has risen from 58 percent in 2012 to 80 percent in 2015, while use of Fixed Price with Economic Price Adjustment has fallen from 41 percent in 2012 to 19 percent in 2015.

For services, which have accounted for 28 percent of contract obligations funded out of the Revolving & Management Funds account since 2012, 77 percent have been structured as fixed price, nearly all as Firm Fixed Price. Of the 19 services contract obligations structured as cost reimbursement since 2012, Cost Plus Fixed Fee accounted for 81 percent, notably higher than the 66 percent of cost reimbursement services contracts for DoD overall. From 2012–2015, use of Cost No Free (from 12 percent to 5 percent) and Cost Plus Award Fee (from 7 percent to 1 percent) have both declined sharply, while use of Cost Plus Incentive (from 4 percent to 11 percent) has risen significantly.

Military Sales Program—Rise of Fixed Price Incentive Use Points to Success of Policy Guidance, But Increased Use of Cost Plus Award Fee Defies Overall Trend

Since 2012, 74 percent of the contract obligations funded out of the Military Sales Program account have been for products, almost entirely structured as fixed price and Firm Fixed Price. Between 2012 and 2015, however, the share of fixed price Military Sales Program products contract obligations structured as Firm Fixed Price has fallen from 97 percent to 80 percent, while the share structured as Fixed Price Incentive has risen from 0 percent to 19 percent, with most of the increase coming in 2014 and 2015. This again can be seen as evidence that policy guidance promoting the use of incentive fee contract types.

Services have accounted for 24 percent of the contract obligations funded out of the Military Sales Program account since 2012, with two-thirds structured as fixed price, roughly in line with the rate for overall DoD services. Almost all of those fixed price contracts are structured as Firm Fixed Price. For the 33 percent of services contracts structured as cost reimbursement, 65 percent were Cost Plus Fixed Fee, while 26 percent were Cost Plus Award Fee. Interestingly, the share structured as Cost Plus Award Fee rose from 13 percent in 2012 to 47 percent in 2015, even as the use of Cost Plus Award Fee declined dramatically for DoD services overall. Over that same period, the use of Cost Plus Fixed Fee fell from 77 percent of all cost reimbursement contract obligations to 42 percent.

5.4. Services Contracting Policy Changes

Over the past year, DoD has made significant changes in how it contracts for services. The following section explores the organizational changes DoD made for services contracting, and developments in the use of contract vehicles for services contracts.

5.4.1. DoD Organizational Changes

Following up on Better Buying Power 3.0's promise of continual departmental improvements and updates, in January 2016 the Department of Defense announced significant changes in the way they would view and acquire services through DoD Instruction 5000.74.

In DoD Instruction 5000.74, the Defense Department laid out a path of standardization for services acquisition through the development of S-CATS, or Services Acquisition Categories. Similar to ACATs (Acquisition Categories), S-CATs have different thresholds and decision authorities. S-CATs also greatly increase the flexibility of acquiring services for the department while maintaining oversight on acquisition programs. S-CATs range from contracts worth $10 million to those worth over $1 billion.[83]

In association with the S-CAT management approach, DoD has designated positions to oversee services acquisition called Functional Domain Experts (FDEs) in order to ensure that those making decisions have substantial experience in each service function. These experts were designated by the USD(AT&L) Frank Kendall.[84] This modification is to ensure that effective and knowledgeable oversight occurs within the acquisition of services. The FDEs provide further management in order to streamline planning and execution, record best practices, and reduce costs across their domain. FDEs will also be able to make policy recommendations for the way forward.[85] To support this process, "Component Level Leads (CLLs) will be appointed by Component heads to assist the FDE in actively overseeing the life-cycle process of contracted services acquisitions."[86]

The S-CAT process also includes establishing Services Requirement Review Boards (SRRBs). These boards will be used to ensure that services contracts meet minimum needs, as well as identifying unneeded requirements and strengthening higher-priority requirements. Various "tripwires" will indicate SRRB intervention, such as services contracts that are worth $10 million or more[87] will automatically be "reviewed, validated, and approved, verifying need and appropriate level of service."[88] This aids in finding areas to reduce both costs and redundancies, while also potentially leading to the decline or elimination of the service in order to fund higher-priority requirements.

SRRBs are also intended to increase visibility of and collaboration on requirements across all stakeholders in an acquisition. This increased collaboration across stakeholders will strengthen collaboration on key strategy decisions that will optimize and make acquisitions more efficient. Overall, SRRBs and the various "tripwires," which also include "labor rates and performance, bridge contracts, use of subcontractors, single-bid procurements and best-value, source-selection premiums," are estimated to save DoD $3 billion over the next five years.[89]

Lastly, DoD is altering data collection and reporting requirements for service contracts in order to increase transparency. This data must now include: total price or total estimated value, total dollar amount obligated, type of contract, whether the contract was performance-based or not, the agency that made the award, the extent of competition,

---

[83] Tony Bertuca, "Pentagon releases landmark services acquisition policy," *Inside Defense*, January 6, 2016, https://insidedefense.com/inside-pentagon/pentagon-releases-landmark-services-acquisition-policy.

[84] Services Acquisition (SA), "Functional Domain Experts: The Role of FDEs in Services Acquisition," http://www.acq.osd.mil/dpap/sa/portfolios_fde.html.

[85] Scott Maucione, "DoD's long-awaited policy streamlines services contracting," *Federal News Radio*, January 5, 2016, http://federalnewsradio.com/acquisition/2016/01/dod-unveils-long-awaited-service-contract-policy/

[86] U.S. Department of Defense, "Instruction, Number 5000.74," January 5, 2016, 3, http://www.dtic.mil/whs/directives/corres/pdf/500074p.pdf.

[87] U.S. Department of Defense, "Instruction, Number 5000.74," January 5, 2016, 22, http://www.dtic.mil/whs/directives/corres/pdf/500074p.pdf.

[88] Ibid., 18.

[89] U.S. Department of Defense, "Instruction, Number 5000.74," January 5, 2016, https://insidedefense.com/insider/dod-services-contract-chief-considers-new-small-biz-policy.

whether the contract award was made to a small business or not, mission to be performed by the contractor, the contracting organization, whether or not the contract is a personal services contract, other contracts that are closely associated, funding source by appropriation and agency, the first fiscal year the contract appeared in ICS requirements, and direct labor hours and associated costs.[90] This modification will greatly aid coordination, transparency, and accountability in services contracting.

A Government Accountability Office (GAO) report released in February 2016, titled "Improved Use of Available Data Needed to Better Manage and Forecast Service Contract Requirements,"[91] addressed the serious lack of available and useful data in DoD's services contracting. These recent alterations on data collection described in DoD 5000.74 are one step toward clarifying and streamlining the contracting process. Often, services contracting data is either unavailable or unusable, which makes tracking trends or making predictions almost impossible. The GAO report recommended that Congress take a leading role on amending DoD reporting requirements to include "information on estimated services contract spending"[92] and that without this reporting requirement, Congress cannot successfully conduct oversight on services contracting within the Department of Defense.

The Defense Department has recognized some of the shortcomings of its current services acquisition system, which has been an area of rapid growth within DoD's budget. In fact, in FY2015, services acquisitions accounted for 53 percent of DoD's budget, totaling at around $143.7 billion. Although DoDI 5000.74 is one step to making these needed alterations, further changes are needed to ensure services are acquired fairly and efficiently across the Department.

5.4.2. Developments in Contract Vehicles for Services

In the acquisition of commercial services, DoD and other government departments such as DHS often rely on the General Services Administration (GSA), which "facilitates the federal government's purchase of high-quality, low-cost goods and services from quality commercial vendors."[93] Open to several federal agencies, GSA operates a vehicle used for services acquisition called OASIS (One Acquisition Solution for Integrated Services). Both OASIS and OASIS SB (Small Business) are multiple award contracts with indefinite length, which are exclusively used for services acquisitions. OASIS serves as a one-stop shop for services acquisition for government agencies and military services. The OASIS process was created to make acquisition of services easier and more efficient for government agencies—and in this respect, the vehicle has been relatively successful.

However, many businesses have concerns with the structure and implementation of these contract vehicles. OASIS has proved a difficult venue for small diverse businesses looking to

---

[90]U.S. Department of Defense, "Instruction, Number 5000.74," January 5, 2016, 26–27, http://www.dtic.mil/whs/directives/corres/pdf/500074p.pdf.

[91] United States Government Accountability Office, "DOD Service Acquisition: Improved Use of Available Data Needed to Better Manage and Forecast Service Contract Requirements," February 2016, http://www.gao.gov/assets/680/675276.pdf.

[92] Ibid., 32.

[93] GSA, "Background and History," https://www.gsa.gov/portal/category/21354.

become a part of the system. A major industry association, The National Defense Industrial Association, in its in-house magazine *National Defense*, interviewed several executives from services companies who believe that high-burden vendor qualification requirements in OASIS result in an approach that "is unfairly stacked in favor of larger firms and penalizes small shops."[94] They cite the expensive sophisticated accounting system preferred by DoD auditors, specialized expensive certifications, and the lack of extensive records as prime contractors. By contrast, task orders under the Air Force's Contracted Advisory and Assistance Services (CAAS) Indefinite Delivery/Indefinite Quantity (IDIQ) contracts have been awarded regularly to small businesses. However, of the current 13 CAAS IDIQ vendors, only one vendor can qualify to compete in the OASIS system. According to some, only "big small businesses" will be able to qualify for OASIS contracting opportunities.[95]

Bid protests have also become an issue with the OASIS service. According to analysis, bid protests have increased 80 percent overall, and with OASIS this trend of significant numbers of bid protests has continued. Though the vehicle has made contracting easier on governmental agencies, the agencies will have to build in additional time in their procurement processes to handle any bid protests that might arise. In the case of OASIS, the government prevailed in each protest, but it still cost the organizations many delays.[96]

GSA's OASIS is similar to the Navy's multiple-award contract (MAC) vehicle, Seaport-e, and one can assume that the growth of Seaport-e spurred GSA to release and promote OASIS. The increasing popularity and growth of these mega-MACs for commercial and professional services, such as OASIS and Seaport-e, is increasing their share of the DoD services market, and hence their effect on DoD services contracting policy.

5.5. Contract Obligations by Effective Competition

In prior CSIS reports on defense acquisition, the study team has noted that, despite concentrated efforts by top policymakers to encourage increases in competition for defense contract obligations, the needle has barely moved in recent years. But while competition rates for DoD contracts overall have been steady, there have been notable changes within the contracting portfolios of the major DoD components; in particular, the decline in competition for Air Force services contract obligations previously identified by CSIS has continued in 2015, and may be more dramatic than previously assumed. Figure 5-7 shows the stability in the level of competition for overall DoD contract obligations.

---

[94] Sandra I. Erwin, "Small Businesses Allege Unfair Contracting Practices in Professional Services Market," *National Defense*, August 1, 2016, http://www.nationaldefensemagazine.org/blog/Lists/Posts/Post.aspx?ID=2262.
[95] Ibid.
[96] Frank Konkel, "GSA Officials on Increased Bid Protests: 'This is how it's going to be,'" *Nextgov*, August 2, 2016, http://m.nextgov.com/technology-news/2016/08/gsa-officials-increased-bid-protests-how-its-going-be/130417/.

Figure 5-7: Level of Competition for Defense Contract Obligations, 2000–2015

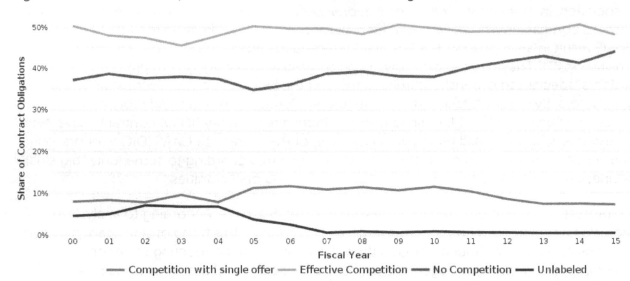

The rate of effective competition[97] for DoD contract obligations has remained between 48 percent and 51 percent in every year between 2008 and 2015. There has not been any significant shift in the quality of that competition, either: the numbers of offers received for effectively competed defense contracts have similarly been fairly static since 2008. The only significant change during the period is a decline in single-offer competition, from 11 percent in 2008 to 7 percent in 2015, in line with policy guidance designed to reduce instances of single-offer competition; that rate has been stable since 2013, however, so progress on that issue appears to have stalled.

---

[97] CSIS defines "effective competition" as competitively sourced contracts receiving at least two offers. This intentionally excludes competitively sourced contracts that receive only one offer; CSIS believes that many of these contracts would have been more appropriately classified as sole-source, and that in any case, DoD is less likely to receive the benefits of competition when there is only one offeror.

Figure 5-8: Overall DoD Platform Portfolio Categories by Rate of Effective Competition

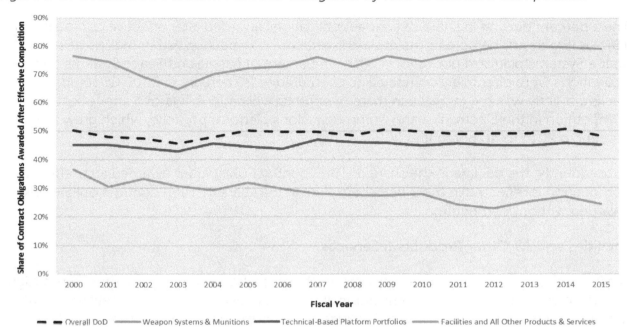

Source: FPDS; CSIS analysis.

Figure 5-8 shows the share of overall defense contract obligations by level of effective competition between 2000 and 2015. After a one-year rise in 2014, the rate of effective competition for each of the platform portfolio categories fell in 2015.

In 2015, just 24 percent of Weapon Systems and Munitions were awarded after effective competition as compared to 27 percent in the previous year. For Technical-Based Platform Portfolio Portfolios (TBPP), the share of contract obligations awarded following effective competition fell from 46 percent in 2014 to 45 percent in 2015. While still overwhelmingly competitive, the share of effectively competed "Facilities and All Other Products & Services" contract obligations fell 1 percent to 79 percent in 2015.

Weapon Systems and Munitions

As the rate of effective competition for the Weapon Systems and Munitions platform portfolio category fell 3 percent in 2015, there were differing trends within each of the different platform portfolios. In 2015, the rate of effective competition fell within the Aircraft and Drones (-4 percent), Land Vehicles (-6 percent), and Weapons and Ammunition (-7 percent). Meanwhile, the rate of effective competition within the Ships and Submarines portfolio increased 6 percent, going from 37 percent in 2014 to 42 percent in 2015.

As the rate of effective competition for Weapon Systems and Munitions fell, the share of contract obligations awarded after no competition increased. In 2015, the share of contract obligations for Weapon Systems and Munition rose to 71 percent from 69 percent in 2014. Similar increases in the share of contract obligations awarded after no competition occurred in the Aircraft and Drones (3 percent), Land Vehicles (8 percent), and Weapons and Ammunition (4 percent) platform portfolios.

Technical-Based Platform Portfolio Portfolios

The 1 percent decline in share of contract obligations awarded after effective competition for TBPP is largely attributable to the decline in effective competition within the Missile and Space Systems platform portfolio. In 2014, 24 percent of Missile and Space Systems contract obligations were effectively competed as compared to 20 percent in 2015. Offsetting this decline slightly was the increase in share of contract obligations awarded after effective competition in the Electronics and Communications platform portfolio, which grew from 44 percent in 2014 to 46 percent in 2015.

Subsequently, the decline in the share of TBPB contract obligations awarded after effective competition is reflected in the 2 percent growth in the share of TBPP contract obligations awarded without competition.

Facilities and All Other Products & Services

While the Facilities and All Other Products & Services platform portfolio category saw a slight decline in the rate of effective competition, it remains overwhelmingly competitive. In 2015, the rate of effective competition for Facilities and Construction remained stable at 75 percent in 2015. The share of contract obligations awarded after effective competition for Other Services fell 3 percent, going from 81 percent to 78 percent. Finally, Other Products remained the most competitive platform portfolio increased the share of contract obligations awarded after effective competition from 85 percent in 2014 to 87 percent in 2015.

6. From Whom Is DoD Buying

It is impossible to discuss trends in DoD contracting without a parallel discussion of the state of the defense industrial base. Increasingly, and for the foreseeable future, DoD depends upon private-sector vendors for platforms, equipment, and supplies to support operations and readiness, and to innovate in order to maintain DoD's technological advantage over current and potential adversaries. While DoD's primary responsibilities are to provide warfighters with the tools and support they need to perform their missions, and to be good stewards of the taxpayers' money, DoD must also pay attention to the impact of conditions and decisions on the health of the defense industrial base.

This chapter seeks to analyze the impact of the budget drawdown on the defense of the industrial base in two ways. The first section looks at changes in the composition of the defense industrial base, as measured by the share of contract obligations going to different size categories of vendors. The second section examines the top vendors in the different market areas of the defense industrial base, and looks at changes in the concentration of those market areas within that top-tier of prime vendors.

6.1. Changes in the Composition of the Defense Industrial Base

Given the dramatic decline in DoD contract obligations during the current budget drawdown, it would be logical to assume that there would be a similarly dramatic effect on the composition of the defense industrial base. Yet while there has been a notable increase in the share of DoD contract dollars obligated to small vendors, the primary changes have been seen not to the overall defense-contracting marketplace, but to specific segments of that marketplace.

To evaluate this impact, CSIS looks at DoD contract obligations to different size categories of vendors: Small, Medium, Large, and the Big 5. Small is defined by the government's classification, with a couple of adjustments that leave CSIS's small business participation shares consistently 2–4 percentage points below what DoD reports. Large is defined as any vendor with over $3 billion in annual revenue from all sources, not just government contracting. A Medium vendor is any vendor that is neither small nor large. And the Big 5 vendors, separated out from large, are vendors that are consistently and by far the largest players in the defense contracting market: Lockheed Martin, Boeing, Northrop Grumman, General Dynamics, and Raytheon. Figure 6-1 shows the composition of the overall defense industrial base.

Figure 6-1:Defense Contract Obligations by Size of Vendor, 2000–2015

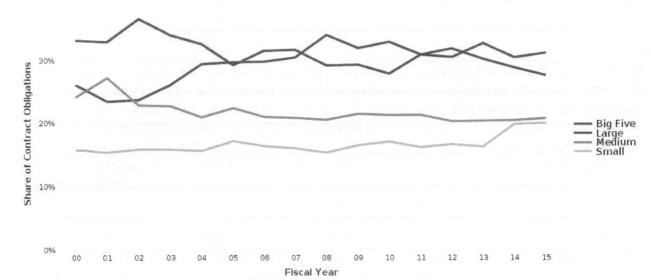

Source: FPDS; CSIS analysis.

Despite the massive decline in DoD contract obligations since 2009, the composition of the defense industrial base has been relatively stable in the 2009–2015 period. Medium vendors have accounted for between 20 percent and 22 percent of overall DoD contract obligations in every year during the period, while Large vendors have accounted for between 30 percent and 34 percent throughout. The Big 5 vendors have seen similar stability, accounting for between 27 percent and 31 percent of DoD contract obligations in each year since 2009. The only notable change has occurred with Small vendors, which have actually increased their share in the last two years, from 16 percent in 2009–2013 to 19 percent in 2014 and 2015. The primary driver of this increase is the rise in small business contract obligations between 2013 and 2014, after the tremendous decline in 2013 in the wake of sequestration. As overall DoD contract obligations declined by 9 percent in 2014, and every other size category saw declines of that magnitude or greater, contract obligations to Small vendors rose by 11 percent.

In 2015, DoD contract obligations to Small vendors declined by 4 percent, roughly in line with the overall decline in DoD contract obligations. Nonetheless, Small is the only category of vendors for which 2015 contract obligations are higher than they were in 2013, which accounts for the increase in share. This data can be seen as a victory for policies that promote Small business participation: despite the pressures of the budget drawdown, Small vendors have managed to not just maintain their place in the defense contracting marketplace, but increase it.

The following sections will examine trends in the composition of the defense industrial base for the different areas of the defense contracting marketplace, as well as those major DoD components with notable trends in recent years.

6.1.1. Research and Development – Massive Decline in Big 5 Market Share Continues in 2015

Figure 6-2: R&D Contract Obligations by Size of Vendor, 2000–2015

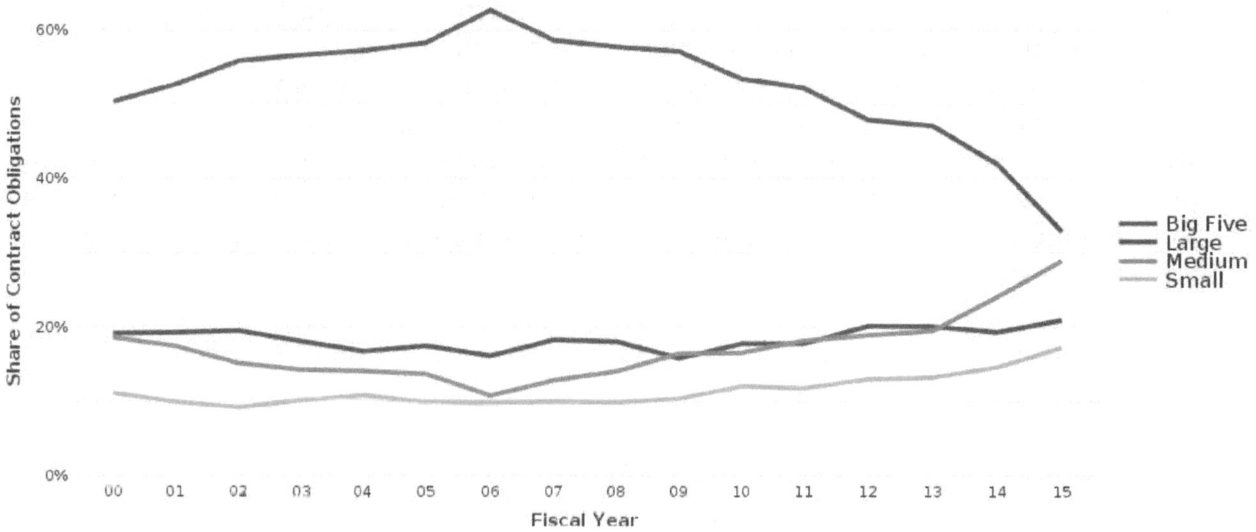

Source: FPDS; CSIS analysis.

Of the three areas of the defense-contracting marketplace, R&D has seen by far the most dramatic shift in the composition of the supporting industrial base. In 2009, the Big 5 vendors accounted for 57 percent of DoD R&D contract obligations. This dominance is not surprising—the largest development programs, for major weapons systems, are disproportionately performed by the Big 5 vendors. But since 2009, the Big 5 share of the DoD R&D contracting market has declined to 33 percent, by far the lowest share in the 2000–2015 period. This is particularly notable because of the massive decline in DoD R&D contract obligations since 2009; overall, the Big 5 control roughly two-fifths less of a market that is less than half the size it was in 2009.

The primary driver of this decline is the now seven-year trough in DoD's development pipeline for major weapons systems that was discussed in Chapter 2. With a number of major development programs either maturing into production or getting canceled, and a dearth of new large development programs starting up, the high-value defense R&D contracting marketplace has shrunk significantly. This trend holds true within all three of the military services, as well as for MDA.

As for the other vendor size categories, the share of DoD R&D contract obligations going to Small vendors has risen from 10 percent to 17 percent since 2009, the share going to Medium vendors has grown from 16 percent to 29 percent, and the share going to Large vendors has risen from 16 percent to 21 percent.

The data for FY2016 shows a continuation of this trend: the share of DoD R&D contract obligations going to the Big 5 vendors continued to fall, from 33 percent to 29 percent, while the share going to Small vendors increased from 17 percent to 19 percent. The continued decline in R&D market share for the Big 5 vendors derives largely from the Army, where the share of R&D contract obligations going to the Big 5 fell from 18 percent in 2015 to just 5 percent in 2016.

As noted in Chapter 2, the trough in DoD's major weapons systems development pipeline is likely to persist in the near term, although how long it will continue varies among the major DoD components. With the B-21 bomber program projected to see funding ramp up over the next few years, the Air Force is likely to be the first of the military services to rise out of the trough. The Navy's Columbia-class ballistic missile submarine program is also in the development pipeline, with development funding likely to ramp up in the next few years now that the program has received Milestone B certification. Meanwhile, the Army is in a far more dire position: given persistent budgetary challenges and continued uncertainty about future missions and required capabilities, it seems unlikely that the Army will emerge from the trough in the near term.

6.1.2. Services—Relatively Stability Despite a Continuing Wave of M&A Activity

Figure 6-3: Services Contract Obligations by Size of Vendor, 2000–2015

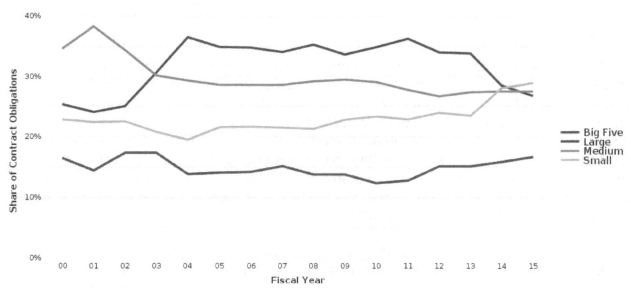

Source: FPDS; CSIS analysis.

Since 2009, there have been minor shifts in the composition in the DoD services industrial base. The share of DoD services contract obligations going to Medium vendors has fallen from 29 percent to 26 percent, and the share going to Large vendors has fallen from 37 percent to 33 percent. Meanwhile, the share going to the Big 5 vendors has risen from 11 percent to 15 percent, and the share going to Small vendors has risen from 21 percent to 26 percent. The increase in Small vendor share was particularly pronounced within three of the five categories of services: FRS&C (32 percent to 39 percent), ICT (32 percent to 39 percent), and PAMS (18 percent to 28 percent) all saw significant increases between 2009 to 2015, to levels well above the Small vendor share of DoD contract obligations overall.

This relative stability over the period, particularly over the last two years, may be somewhat surprising given the wave of M&A activity within the government services sector in recent years. A nonexhaustive list of recent major M&A activity within the sector includes:

- The spin-off of ITT's defense business into Exelis, which was then subsequently purchased by the Harris Corporation.

- The spin-off of Computer Sciences Corporation's government services business, which then merged with SRA international to form CSRA.

- The spin-off of L3 Communication's government services business into Engility, which then acquired TASC, Inc.

- The recently completed spin-off of Lockheed Martin IT services business, which is expected to merge with Leidos, which was itself a spin-off from SAIC.

- KBR's recent acquisitions of Wyle and of Honeywell's government services business.

The overall trend with the government services market is twofold: first, a trend of diversified vendors divesting their government services business units (particularly in government IT services); and second, of government services-focused vendors merging with or acquiring other vendors to increase market share and access to markets/sectors.

These changes, however, have not yet been reflected in the data on the composition of the DoD services industrial base, even when looking at the data for FY2016. This is partially the result of how recently a number of the larger mergers and spin-offs have occurred, including ones yet to be formally completed, but also a factor of the precise nature of the activity relative to this method of analysis. When CSC, a Large vendor, spun off its government services business unit, the resulting company was also Large. Similarly, ITT, Exelis, and Harris Corp are all Large vendors, so the activity between them didn't cause contract obligations to shift between size categories. Much of other M&A activity that has occurred in the government services sector has been between Medium vendors which, even when combined, do not have enough annual revenue to qualify as Large.

As this wave of M&A continues, CSIS hopes to continue to explore alternate methods to quantify the scope and impact of this shift in the composition of the DoD services industrial base.

6.1.3. Products—A Broad-Based Shift toward the Big 5 Vendors

Figure 6-4: Products Contract Obligations by Size of Vendor, 2000–2015

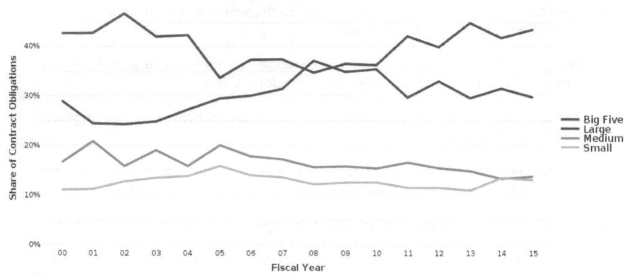

Source: FPDS; CSIS analysis.

Unlike for services and R&D, there has been no increase in the share of DoD products contract obligations going to Small vendors—the share has remained between 11 percent and 13 percent in every year between 2009 and 2015. Medium vendors have seen similar stability, with their share of the DoD products contracting market fluctuating between 13 percent and 17 percent. There has, however, been a somewhat notable shift in share between Large vendors and the Big 5: the share going to Large vendors has fallen from 35 percent in 2009 to 30 percent in 2015 while the share going to the Big 5 has risen from 36 percent to 43 percent. No one category of products was the driver of this trends; rather, there were moderate increases in the share going to the Big 5 and moderate declines in the share going to Large vendors in a number of product categories. This is despite a large shift away from the Big 5 to Large in Ships due to the spin-off of Northrop Grumman's shipbuilding business into Huntington Ingalls Industries back in 2011.

6.1.4. Army—Small Vendors See Significant Gains

Figure 6-5: Army Contract Obligations by Size of Vendor, 2000–2015

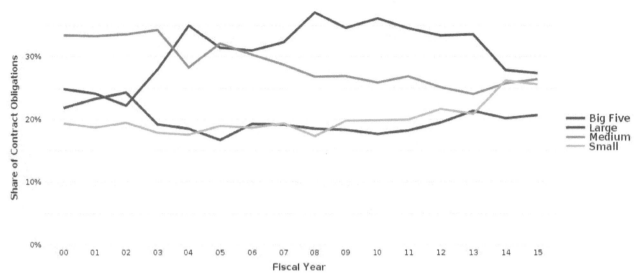

Source: FPDS; CSIS analysis.

As with DoD overall, both Medium and Big 5 vendors have seen relative stability in their shares of overall Army contract obligations since 2009. Medium vendors have accounted for between 24 percent and 27 percent of Army contract obligations in each year between 2009 and 2015, while the Big 5 have accounted for between 18 percent and 21 percent. The share of Army contract obligations awarded to Large vendors fell from 35 percent to 27 percent between 2009 and 2015, while the share going to Small vendors rose from 20 percent to 26 percent.

Looking at the different areas of the Army contracting industrial base, there were significant increases in the shares of both Army services and Army R&D contracting obligations going to Small vendors. The share of Army services going to Small vendors rose from 24 percent in 2009 to 33 percent in 2015, while the share of Army R&D going to Small vendors rose from 18 percent to 29 percent; the share for Small vendors increased to 34 percent in 2016. The shift away from Large vendors, meanwhile, was concentrated within Army products and Army services: the share of Army products contract obligations going to Large vendors fell from 39 percent in 2009 to 32 percent in 2015, while the share of Army services going to Large vendors fell from 35 percent to 25 percent.

6.1.5. Navy—Stability Overall, But a Shift toward Small Vendors in Services and R&D

Figure 6-6: Navy Contract Obligations by Size of Vendor, 2000–2015

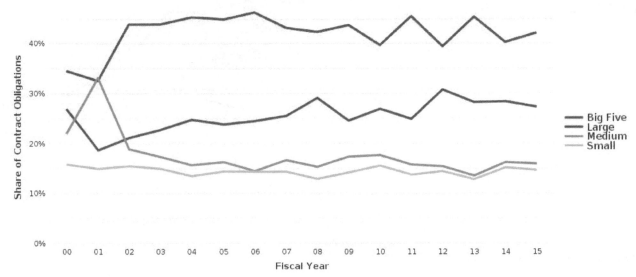

Source: FPDS; CSIS analysis.

Unlike the Army, the Navy has seen relative stability across all size categories of vendors. The share of Navy contract obligations going to Small vendors has remained between 13 percent and 16 percent throughout the 2009–2015 period, while the share going to Medium vendors has remained between 15 percent and 18 percent. Large vendors have seen more fluctuation in market share, but have remained between 25 percent and 28 percent in all but one year from 2009–2015 (31 percent in 2012.) The share going to the Big 5 vendors has fluctuated similarly, remaining between 39 percent and 45 percent in each year, with the year-to-year fluctuations primarily driven by the timing of large F-35 purchases.

The Navy has actually seen moderate increases in the share of both services and R&D contract obligations going to Small vendors: the share of Navy R&D going to Small vendors rose from 9 percent in 2009 to 17 percent in 2015, while the share of Navy R&D going to Small vendors rose from 26 percent in 2009 to 32 percent in 2015. Meanwhile, the share of Navy products contract obligations going to Small vendors fell from 8 percent to 5 percent over the 2009–2015 period, and since products account for over half of overall Navy contract obligations, the overall Navy data does not show a significant shift towards Small vendors.

6.1.6. Air Force—R&D Contract Obligations Shift toward Small and Medium Vendors

Figure 6-7: Air Force Contract Obligations by Size of Vendor, 2000–2015

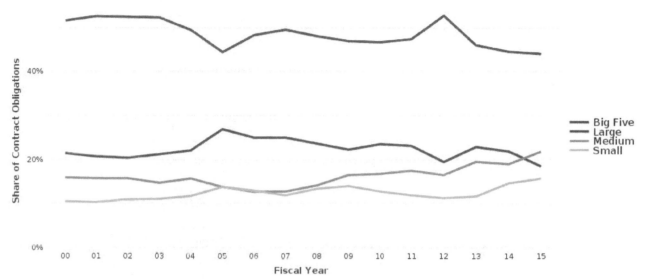

Source: FPDS; CSIS analysis.

The composition of the industrial base supporting Air Force contracting has shifted moderately across the four size categories of vendors. The share of Air Force contract obligations going to small vendors rose from 14 percent to 16 percent between 2009 and 2015, while the share going to Medium vendors rose from 17 percent to 22 percent. Meanwhile, Large vendors fell as a share of Air Force contract obligations from 23 percent to 18 percent, and the Big 5 vendors declined from 47 percent of overall Air Force contract obligations to 44 percent.

Unlike the Navy, the Air Force actually saw a small decline in the share of services contract obligations going to Small vendors, but the share of R&D going to Small vendors nearly doubled, from 8 percent in 2009 to 14 percent in 2015. The data show a similar pattern for Medium vendors: there was only a slight increase in the share of Air Force services contract obligations going to Medium vendors, but within R&D the Medium share increased from 30 percent in 2009 to 46 percent in 2015.

6.1.7. Defense Logistics Agency—Small Vendors See Significant Gains in DLA Products Contracting

Figure 6-8: DLA Contract Obligations by Size of Vendor, 2000–2015

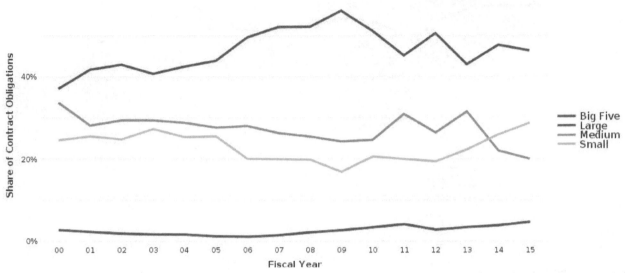

Source: FPDS; CSIS analysis.

There has been a significant increase in the share of DLA contract obligations going to Small vendors since 2009, as that share rose from 17 percent in 2009 to 29 percent in 2015, the highest of any major DoD component in any year during the 2000–2015 period. This increase was broad-based between 2009 and 2015, across both areas within DLA's contracting portfolio with significant contract obligations: Small vendors' share of DLA products rose from 17 percent to 29 percent, and their share of DLA services rose from 21 percent to 37 percent.

6.2. The Present and Future Consolidation of Defense Industry

As part of analyzing the composition of the defense industrial base, CSIS tracks the top vendors for DoD overall, as well as for the different market areas and within the major DoD components. In previous years, the study team has looked at 10-year periods, but given the profound changes in DoD contracting during the current budget drawdown, this analysis focuses on a comparison between 2009, the peak year before the drawdown, and 2015. Table 1 shows the top 20 vendors for overall DoD contract obligations in 2009 and 2015.

Table 1: Top 20 Defense Vendors, 2009 and 2015

Top 20 Contractors in 2009	Obligations in 2015 Millions	2008 Rank	Top 20 Contractors in 2015	Obligations in 2015 Millions	2014 Rank
Lockheed Martin	34,611	1	Lockheed Martin	29,243	1
Boeing	23,083	2	Boeing	14,471	2
Northrop Grumman	20,052	3	Raytheon	11,985	4
General Dynamics	17,905	5	General Dynamics	11,250	3
Raytheon	17,827	6	Northrop Grumman	9,622	5
Top 5 Total	**113,477**		**Top 5 Total**	**76,571**	
BAE Systems	8,132	4	United Technologies	6,585	6
L3 Communications	7,941	8	L3 Communications	5,169	7
United Technologies	7,551	7	BAE Systems	4,728	8
Oshkosh	6,912	30	Humana	3,553	10
SAIC	5,564	12	Bechtel	2,977	14
KBR	5,130	9	Huntington Ingalls	2,799	9
ITT	4,170	13	Health Net	2,765	12
Humana	3,796	16	SAIC	2,511	11
Computer Sciences Corp.	3,598	14	UnitedHealth Group	2,505	13
General Electric	3,314	15	General Atomics	2,304	21
Health Net	3,133	19	McKesson	2,143	20
AM General	3,007	11	Bell-Boeing Joint Project Office*	2,043	18
TriWest Healthcare	2,955	20	AmerisourceBergen	1,843	25
Bell-Boeing Joint Project Office*	2,947	18	Booz Allen Hamilton	1,802	16
Bechtel	2,711	24	United Launch Alliance*	1,723	15
Top 20 Total	**184,340**		**Top 20 Total**	**122,024**	
Overall DoD Total	**420,722**		**Overall DoD Total**	**272,286**	

Source: FPDS; CSIS analysis. (Joint venture).*

This table, and the ones like it that follow in this section, show data for prime contract obligations, as FPDS only includes prime contract data. There is a separate subcontract database that CSIS is currently developing tools to analyze, but those efforts are still ongoing. The chart shows the top vendors in 2009 and 2015, their total obligations in those years, and their rank in the previous year.

Between 2009 and 2015, the top 5 vendors for DoD contract obligations have stayed the same, though the order has changed: Northrop Grumman, which was the 3rd-largest DoD vendor in 2009, is 5th in 2015, largely due to the spinoff of their shipbuilding business into Huntington Ingalls Industries, which itself ranked 11th in 2015. Seven of the top 20 vendors in 2009 are not in the top 20 in 2015: ITT and CSC divested their government services businesses, while Oshkosh, KBR, General Electric, AM General, and TriWest Healthcare have simply declined as players in the overall DoD marketplace, though KBR seems likely to rise going forward due to their recent acquisitions.

In 2009, it took $2.7 billion of contract obligations to make it into the top 20 DoD vendors; by 2015, that threshold had been reduced to $1.7 billion. The data show no significant shift in the concentration of the defense industrial base overall since 2009: the share of total DoD contract obligations going to the top 5 vendors was virtually the same in 2009 (27 percent) as in 2015 (28 percent), and the same is true when looking at the share of overall DoD contracts going to the top 20 vendors in 2009 (44 percent) and 2015 (45 percent).

The following sections look at the top vendors within the three market areas of the defense industrial base.

6.2.1. Top Products Vendors—Defense Products Industrial Base Becomes Increasingly Concentrated during Budget Drawdown

Table 2: Top 20 Defense Products Vendors, 2009 and 2015

Top 20 Contractors in 2009	Obligations in 2015 Millions	2008 Rank	Top 20 Contractors in 2015	Obligations in 2015 Millions	2014 Rank
Lockheed Martin	18,269	1	Lockheed Martin	21,357	1
Boeing	12,941	3	Boeing	10,452	2
General Dynamics	12,650	5	General Dynamics	9,023	3
Raytheon	10,830	6	Raytheon	8,237	4
Northrop Grumman	9,128	4	United Technologies	5,003	5
Top 5 Total	**63,818**		**Top 5 Total**	**54,072**	
Oshkosh	6,810	17	Northrop Grumman	3,895	7
United Technologies	5,631	7	Huntington Ingalls	2,622	6
BAE Systems	5,354	2	Bechtel	2,485	9
AM General	2,980	9	McKesson	2,141	13
L3 Communications	2,971	14	BAE Systems	2,137	8
Bell-Boeing Joint Project Office*	2,860	10	Bell-Boeing Joint Project Office*	2,042	10
General Electric	2,803	11	AmerisourceBergen	1,843	15
BP	2,435	16	L3 Communications	1,724	11
Agility	2,165	15	General Atomics	1,627	23
Royal Dutch Shell	2,093	18	Textron	1,419	16
ITT	1,958	13	Oshkosh	1,344	35
Bahrain Petroleum Company	1,942	26	General Electric	1,205	12
ATK	1,815	19	Atlantic Diving Supply	1,070	19
Textron	1,622	12	ATK	992	18
Navistar	1,456	8	AM General	930	65
Top 20 Total	**108,713**		**Top 20 Total**	**81,549**	
Overall Products Total	**189,541**		**Overall Products Total**	**128,920**	

Source: FPDS; CSIS analysis. (Joint venture).*

The only change to the top 5 DoD products vendors was the decline of Northrop Grumman from 5th to 6th, primarily due to their spin-off of their shipbuilding business into Huntington Ingalls Industries, which ranked 7th in 2015. Replacing them is United Technologies, which was 7th in 2015; with United Technologies sale of their Sikorsky business unit to Lockheed Martin, they are likely to fall out of the top 5 next year. Six companies that were in the top 20 in 2009 were not among the top 20 DoD products vendors in 2015: three fuel suppliers (BP, Royal Dutch Shell, and the Bahrain Petroleum Company), Agility (a Kuwaiti logistics company), ITT, and Navistar. Three companies that were in the top 20 in 2015 were not in the top 20 in 2014: General Atomics (23rd in 2014), Oshkosh (35th in 2014), and AM General (65th in 2014.)

In 2009 the threshold for being among the top 20 DoD products vendors was $1.5 billion in contract obligations; by 2015, the 20th ranked products vendor accounted for slightly more than $900 million. There has been a significant increase in the concentration of the DoD products market among the top 5, with the share of overall DoD products contract obligations going to the top 5 rising from 34 percent in 2009 to 42 percent in 2015. Similarly, the share captured by the top 20 vendors rose from 57 percent in 2009 to 63 percent in 2015.

6.2.2. Top Services Vendors—Notable Turnover among Top 5 Service Vendors Since 2009

Table 3: Top 20 Defense Services Vendors, 2009 and 2015

Top 20 Contractors in 2009	Obligations in 2015 Millions	2008 Rank	Top 20 Contractors in 2015	Obligations in 2015 Millions	2014 Rank
Lockheed Martin	7,136	3	Northrop Grumman	4,126	2
Northrop Grumman	5,716	2	Lockheed Martin	4,117	1
KBR	5,112	1	Humana	3,553	3
L3 Communications	4,578	4	Boeing	3,537	4
SAIC	3,805	7	L3 Communications	3,123	6
Top 5 Total	**26,347**		**Top 5 Total**	**18,456**	
General Dynamics	3,804	5	Health Net	2,765	5
Humana	3,796	8	Raytheon	2,679	10
Computer Sciences Corp.	3,332	6	UnitedHealth Group	2,505	7
Raytheon	3,280	9	BAE Systems	2,194	11
Health Net	3,133	11	Humana	1,913	12
TriWest Healthcare	2,955	12	SAIC	1,826	9
Boeing	2,923	10	United Launch Alliance*	1,715	8
URS	2,487	13	Booz Allen Hamilton	1,298	14
BAE Systems	2,163	14	DynCorp International	1,165	18
ITT	1,821	18	Computer Sciences Corp.	1,149	15
Hensel Phelps	1,804	19	URS	1,077	19
Booz Allen Hamilton	1,696	20	Hewlett-Packard	998	13
FedEx	1,677	15	CACI	993	17
Hewlett-Packard	1,573	148	Vectrus	836	N/A
CACI	1,549	21	Fluor	704	20
Top 20 Total	**64,341**		**Top 20 Total**	**42,273**	
Overall Services Total	**183,674**		**Overall Services Total**	**120,952**	

Source: FPDS; CSIS analysis. (* Joint venture).

There has been a significant shift in the composition of the top 5 defense services vendors between 2009 and 2015. KBR and SAIC, which were in the top 5 in 2009, have dropped out in 2015; KBR is not in the top 20 at all in 2015, and SAIC (which divided its business between the SAIC name and Leidos in September 2013) has declined to 11th. They have been replaced by Humana, which was 7th in 2009, and Boeing, which was 12th in 2009. Only three of the top 20 in 2009 are not in the top 20 in 2015: TriWest Healthcare, ITT, and FedEx. Only one vendor in the top 20 in 2015 was not in the top 20 in 2014: Vectrus, which was spun off from Exelis (which was itself spun off from ITT, and is now merging with Harris Corp.)

In 2009 it took $1.5 billion in contract obligations to be a top 20 services vendor for DoD; in 2015 the 20th-ranked vendor accounted for less than half that figure (roughly $700 million.) As with DoD overall, there has not been a significant change in the concentration of the defense services industrial base: as a share of overall defense services contract obligations, the top 5 was relatively unchanged between 2009 (14 percent) and 2015 (15 percent). Similarly, the top 20 accounted for 35 percent of total defense services contract obligations in both 2009 and 2015.

6.2.3. Top Research and Development Vendors—Massive De-concentration due to Trough in MDAP Development Pipeline

Table 4: Top 20 Defense R&D Vendors, 2009 and 2015

Top 20 Contractors in 2009	Obligations in 2015 Millions	2008 Rank	Top 20 Contractors in 2015	Obligations in 2015 Millions	2014 Rank
Lockheed Martin	9,206	1	Lockheed Martin	3,769	1
Boeing	7,218	2	Northrop Grumman	1,601	2
Northrop Grumman	5,208	3	Raytheon	1,070	3
Raytheon	3,717	4	MIT	966	5
MIT	1,931	9	United Technologies	914	10
Top 5 Total	**27,280**		**Top 5 Total**	**8,320**	
General Dynamics	1,451	5	Aerospace Corp.	840	9
United Technologies	1,184	6	MITRE	751	7
SAIC	907	8	Johns Hopkins APL	733	8
Aerospace Corp.	875	7	Booz Allen Hamilton	499	6
Booz Allen Hamilton	867	10	Boeing	481	4
BAE Systems	615	12	Alion Science & Technology	399	11
Johns Hopkins APL	486	16	BAE Systems	397	12
GE Rolls-Royce Fighter Engine Team*	452	14	Leidos	348	16
L3 Communications	392	15	Wyle Laboratories	331	14
ITT	391	11	L3 Communications	322	15
MITRE	369	19	General Dynamics	315	13
JVYS*	318	18	Georgia Institute of Technology	219	21
Battelle	285	17	CACI	194	20
CACI	272	24	Battelle	191	17
Computer Sciences Corp.	257	22	SAIC	160	18
Top 20 Total	**36,401**		**Top 20 Total**	**14,503**	
Overall R&D Total	**47,506**		**Overall R&D Total**	**22,414**	

Source: FPDS; CSIS analysis. (* - Joint Venture).

Boeing, which was the 2nd-ranked DoD R&D vendor in 2009, has declined to 10th in 2015, after being 4th in 2014; it has been replaced in the top 5 by United Technologies, which was 7th in 2009 and 10th in 2014. Four vendors in the top 20 in 2009 were not in the top 20 in 2015: the GE/Rolls Royce Fighter Engine Team joint venture, ITT, JVYS (a joint venture between Yulista and SES-I), and Computer Sciences Corporation. Only one vendor in the top 20 in 2015 was not in the top 20 in 2014: the Georgia Institute of Technology, which ranked 21st in 2014.

In 2009 the threshold to be among the top 20 DoD R&D vendors was roughly $260 million; in 2015 that threshold was only $160 million. There has been a massive decline in the concentration of the DoD R&D industrial base since 2009: the top 5 share of overall DoD R&D contract obligations has declined from 57 percent to 37 percent, and the top 20 share has fallen from 77 percent to 65 percent. This massive decline in sector concentration is the result of the now seven-year trough in DoD's development pipeline for major weapons systems discussed in the "Research and Development Contracting During the Budget Drawdown" section of Chapter 4. Since the largest vendors disproportionately perform the largest R&D projects, it is not surprising that a dearth of large development programs would drastically reduce the share of R&D going to those large vendors.

6.3. International Joint Development (IJD)[98]

For most of the world, the question is not whether to do international joint development of new capabilities, but when and how, because most countries simply do not have the capacity and funding to go it alone on highly complex and expensive development efforts. Traditionally, the United States has had, and has exercised, the option to go it alone on major development efforts, but even increased U.S. military funding will not eliminate the need to make tradeoffs in the allocation of U.S. research and development funding, and the increasingly globalized technology base will serve as a growing incentive to look abroad for partners in developing cutting-edge technology. Hence it is likely that the United States also will see the need to engage in international joint development going forward, as it has many times historically.

6.3.1. Partnering with a multinational industrial base

When the United States and other countries cooperate to jointly fund, develop, and produce a weapon system, the defense acquisition system buys from an expanded world of vendors. International cooperation on development projects can be beneficial in both the private and public spheres, as well as for both military and civil projects. International joint development projects in defense merit special attention because the barriers to cooperation in that sector are particularly high, even within alliances. While unique combinations of benefits drive each international program, most nations turn to international cooperation in defense acquisition to appease budget pressures and procure advanced programs that they cannot individually afford.

DoD recognizes the value of international joint development programs that include both research funding from, and technology development with, multiple countries. This is especially true in light of the Budget Control Act of 2011, which imposed caps on defense spending concurrent with European defense budget reductions. In reaction to a fiscally constrained environment, January 2012's Defense Strategic Guidance committed DoD, and the United States at large, to strengthening partnership and cooperation with the global community by emphasizing pooling, sharing, and specializing capabilities with partner nations.[99]

DoD's support for international joint development comes with policies that determine when international joint development is and is not appropriate. The International Cooperation in Acquisition, Technology and Logistics Handbook states that when considering the pursuit of an international joint development program, the Milestone Decision Authority must consider whether a program executes "demonstrated best business practices, including a plan for

---

[98] *This material is adapted from a previous CSIS study,* Designing and Managing Successful International Joint Development Programs, *supported by the Naval Postgraduate School Acquisition Research Program under Contract No. HQ0034-12-A-0022-0008. The views expressed in written materials or publications, and/or made by speakers, moderators, and presenters, do not necessarily reflect the official policies of the Naval Postgraduate School nor does mention of trade names, commercial practices, or organizations imply endorsement by the U.S. government.*
[99] United States Department of Defense, *Sustaining U.S. Global Leadership: Priorities for the 21st Century Defense,* 2012.

effective, economical, and efficient management of the international cooperative program."[100] While the value of international joint development programs is recognized, the theoretical basis for best practices in these programs is scarce.[101]

Why take on the complex task of international cooperation?

Why does the U.S. government embark on these organizationally complex international programs? What are their theoretical benefits, and how are they appealing enough to incentivize such ambitious and complicated goals? Maintaining a national competitive edge in the globalized economy, addressing monopolistic and oligopolistic market failures, collective action inefficiencies, financial needs, life-cycle costs, lack of competition, and alliance cohesion are among the many drivers pushing the pursuit of international joint development in defense acquisition.[102] Additionally, there are theoretical benefits from international cooperation that appeal to countries that experience the drivers listed above, which include shared R&D costs, shared risk, improved learning economies, greater economies of scale, lower unit cost of weapons procured, end-products taking advantage of specializations in other countries, and military interoperability.[103]

While these hypothetical benefits merit support for international programs in defense acquisition, there are complications that exist in practice and often prevent international programs from achieving the benefits previously outlined. Single-nation acquisition is hard, and international joint acquisition is harder. Evidence from past programs shows that international programs encourage participants to behave opportunistically, face collective tradeoffs that result in sub-optimal end products for individual nations, and experience competing factors within their structures. These phenomena obstruct international programs from achieving their hypothetical benefits and knowledge of them is key to shaping best practices in the future.

6.3.2. Best Practices in International Joint Development

Nations considering international cooperation in defense acquisition should critically question their incentives to do so. Single-nation acquisition is hard, and international joint acquisition is harder. Therefore, countries considering joint cooperation should ensure that they are positioned to handle the additional challenges associated with international programs. Countries should use a high burden of proof when conducting both a risk and a cost/benefit analysis. They should then ask themselves if the additional risks and costs associated with international cooperative programs are a better option than defense trade or pursuing the program indigenously. The CSIS study from which this section is drawn, "Designing and Managing Successful International Joint Development Programs," created a

---

[100] Office of the Undersecretary of Defense for Acquisition Technology and Logistics, "International Cooperation in Acquisition, Technology and Logistics (IC in AT&L) Handbook," 2012, http://www.acq.osd.mil/ic/Links/IChandbook.pdf.

[101] Ibid.

[102] M.R. De Vore, "The Arms Collaboration Dilemma: Between Principal-Agent Dynamics and Collective Action Problems," *Security Studies*, Vol. 20, 2011, http://www.tandfonline.com/doi/pdf/10.1080/09636412.2011.625763. B. Fitzgerald, B. Greenwalt, S. Grundman, J. Hasik, and R. Rumbaugh,,*International defence industrial co-operation in the post-financial crisis era,* 2014.

[103] De Vore, "The Arms Collaboration Dilemma," 625, 627, 628.

framework tracking 10 characteristics of these projects. Based on that framework, the study identified key questions to aid in evaluating the costs and benefits as well pointing to models that mitigate its challenges.

Considering Operational and Political Factors

When procuring a program with other militaries, a variety of wants and needs compete against each other. This also notably happens among the military services of the United States, where inter-service consensus on common requirements can be just as difficult to achieve as international agreement. What is clear is that successful international joint development programs should satisfy multiple categories of objectives, be they security, political, or economic, for each participant. These objectives need to have strong champions, who need to be capable of working with their international counterparts to overcome domestic constraints. Programs should also have comparatively few dedicated domestic opponents or they will face a magnified risk of breaking apart should major challenges arise.

As a result, some reasons for pursuing international joint development can serve as valuable secondary objectives but are insufficient to serve as primary objectives. Providing competition to existing indigenous systems through international joint development will likely face steady domestic opposition while having only mixed operational support, because the indigenous alternative is always available. International joint development can motivate partners to take the time and effort necessary to navigate export control restrictions. However, seeking to start up a cutting-edge domestic capacity through international joint development may stumble on the dual hurdles of insufficient domestic economic foundation and technology transfer limitations. Similarly, if multiple countries want the political and economic benefits of being the prime contractor, the organizational complexity and economic challenges of international joint development may easily undercut the advantages of working together.

Ensuring that a program serves multiple categories of objectives also mitigates against setting ambitious goals for operational capabilities, technology transfer, or industrial base development. Pushing too hard toward any one of these goals in isolation is likely to undercut other benefits for the project as a whole or for other participants. Thus, great ambitions in any one category typically come at the expense of the project's suitability for international joint development.

Designing component compartmentalization into programs

Compartmentalized workshare distribution is dividing work into discrete, severable design elements that require a minimum of subsequent design integration—a practice exemplified in two cases studied by CSIS: the SM-3 Block IIA ship-based missile defense program involving the United States and Japan and the final, successful iteration of the NATO Advance Ground Surveillance program. A compartmentalized approach can optimize participants' cost/benefit ratios by minimizing integration complexity involving both industry teams and national governments, as well as the risk from technology transfer limitations, while also increasing economic benefits. In a best-case scenario, this ensures successful program outcomes while also optimizing individual country outcomes across a variety of objectives. However, this approach requires partners with the industrial capability to design and build major complex system elements.

Techniques for Mitigating Competing Objectives

Establishing a broad portfolio of international collaborative projects, so that workload can be allocated as part of this larger portfolio, can constitute a best practice. A group of closely allied countries that collaborate often would benefit from practicing the prime-sub model of cooperation across all collaborative programs and also give different countries opportunities to lead as the prime contractor for each collaborative program. Looking at a wider portfolio makes this process easier. Beyond the scope of this paper, there are also a range of other forms of international cooperation that could be considered, from off-the-shelf defense trade to coproduction to international procurement of subcomponents. This bypasses the collective action issue that most collaborative programs face, in which partners attempt to maximize their benefits and deflate their costs.

Joint Development as a Mechanism for Advancing Technological Capabilities

International cooperation in defense acquisition poses various hurdles associated with technology transfer laws. Countries that are global leaders in technology often implement demanding bureaucratic processes when participating in international armaments cooperation. Additionally, countries that seek international cooperation in armaments sometimes do so in hope of receiving industrial spillover benefits that result from technology and information sharing. Historically, countries engaging in international joint development to achieve these objectives often fail to achieve their anticipated spillover benefits due to restrictions mandated by partner country's technology security regime. Such a shortfall threatens the program because partner nations that fail to accrue their expected benefits from industrial spillovers are more likely to defect. Countries that choose to participate as joint development partners are less likely to face high technology transfer hurdles through processes, such as the U.S. International Traffic in Arms Regulations, compared to those simply pursing defense trade.[104]

---

[104] The International Traffic in Arms Regulations enumerates requirements for the export and import of items and information ("defense articles" and "defense services") found on the U.S. Munitions List. The U.S. Department of State Directorate of Defense Trade Controls is responsible for implementing ITAR through the issuing of export licenses.

7. What Are the Defense Components Buying?

This sections examines changes within the contracting portfolios of the major DoD components (Army, Navy, Air Force, DLA, MDA, and Other DoD, which includes all contracting entities not captured by the first five categories), focusing on the period since 2012, the year before the initial impact of sequestration is registered in the data. Each section looks at trends within the different market areas of each component's contracting portfolio, as well as examining competition and industrial base trends for the three military services.

While the data shows that the overall decline in DoD contract obligations may be close to leveling off in 2015, there are notable differences in trends between the major DoD components, and in the particular drivers of those trends. Figure 7-1 shows DoD contract obligations, broken down by major DoD component, from 2000–2015.

Figure 7-1: Defense Contract Obligations by Component, 2000–2015[105]

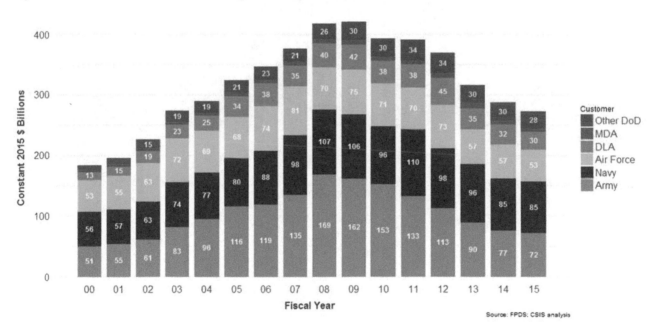

Source: FPDS; CSIS analysis.

Source: FPDS; CSIS analysis.

[105] CSIS has recently discovered that all contract obligations for the F-35 Joint Strike Fighter are categorized as Navy contract obligations in FPDS. This is not a data error, but rather a reflection of the way the contracts and contract management of the program are structured. As a result, FPDS data overstates Navy contract obligations and understates Air Force contract obligations; because this is a consistent issue, however, the study team does not believe it significantly impedes analysis of trends. CSIS is investigating whether and how to properly categorize the Air Force's share of F-35 contract obligations for future analysis.

7.1. Army—Massive Decline Begins to See Bottom in 2015

Of the three military services, the Army has been hit hardest by the declines in DoD contract obligations since 2012, falling by 36 percent over the 2012–2015 period, notably more steeply than overall DoD contract obligations. 2014 saw a 14 percent decline within the Army's contracting portfolio, which was also steeper than the decline for overall DoD, but in 2015, Army contract obligations declined by only 6 percent, which was roughly in line with the overall decline in DoD contract obligations. This may indicate that the steep decline in Army contract obligations since 2009, driven by the winding-down of combat operations in Iraq and Afghanistan and the Army's recent inability to start and sustain major development programs, may be close to reaching its bottom. Figure 7-2 shows Army contract obligations, broken down by area, from 2000–2015.

Figure 7-2: Army Contract Obligations by Area, 2000–2015

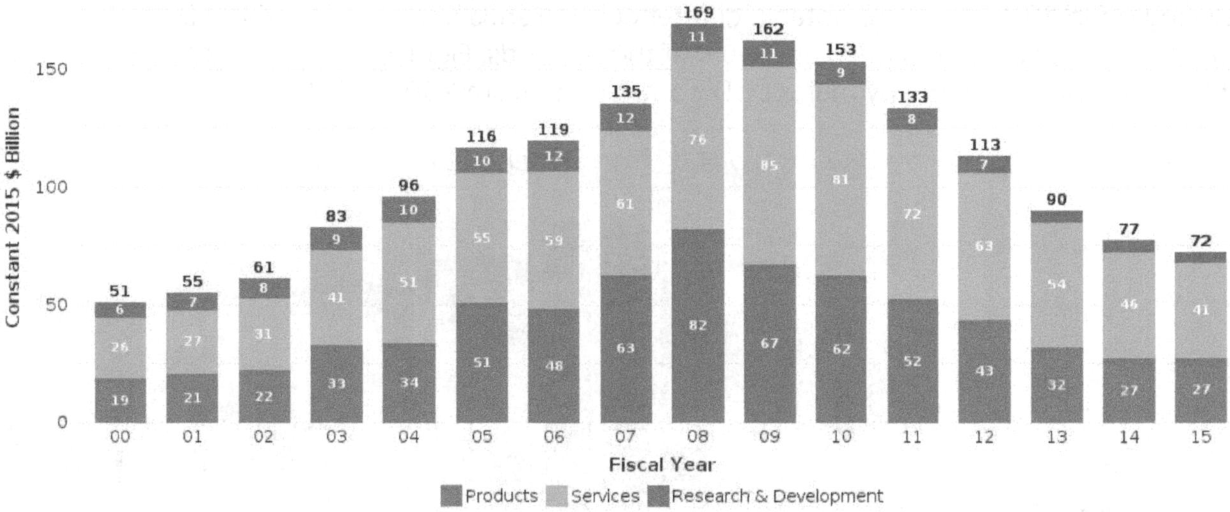

Source: FPDS; CSIS analysis.

Figure 7-3: Army Products Contract Obligations by Category, 2000–2015

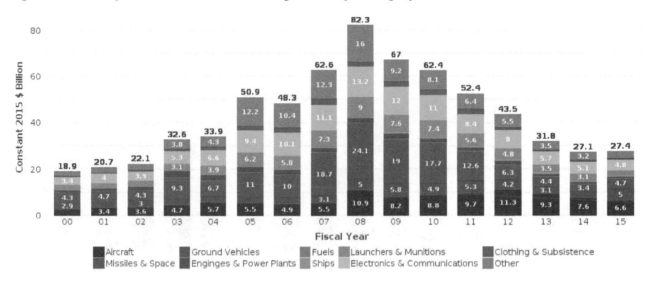

Source: FPDS; CSIS analysis.

Since 2012, Army products contract obligations have declined by 37 percent, roughly in line with the overall decline for Army contracts, but significantly steeper than the decline in overall DoD products over that same period. 2014 saw a 15 percent decline, which was also in line with the overall Army decline in that year, but in 2015, Army contract obligations actually increased by 1 percent, in a year when overall Army contracts were declining moderately.

Looking at the different categories of products within the Army's products contracting portfolio, obligations for Missiles & Space increased by a massive 85 percent between 2014 and 2015, from $2.7 billion to $5 billion, with almost all of that increase related to the Patriot and MEADS missile defense programs. Obligations for Ground Vehicles also increased strongly (38 percent) from 2014–2015, with an almost $500 million increase in obligations related to the Stryker program offset by a $1.6 billion increase in obligations for "Trucks and Truck Tractors, Wheeled," which were not labeled by program in FPDS.

Several categories of products also saw significant declines in 2015. Army contract obligations for Aircraft declined by 14 percent, with a $1 billion decline in obligations related to the AH-64A Apache attack helicopter program and a $1.1 billion decline in obligations for the Scout helicopter program, after a one-year spike in 2014. These declines outweighed significant increases in obligations for "Airframe Structural Components" related to the UH-72A helicopter program and various drones not labeled by program in FPDS. Army obligations for Launchers & Munitions declined by 23 percent, driven by a nearly $300 million decline in obligations for "Fire Control Systems, Complete."

Among the other Army products contract categories with significant obligations, contract obligations for Clothing & Subsistence declined by 26 percent in 2015, Electronics &

Communications declined by 6 percent, Engines & Power Plants declined by 33 percent, and "Other" products decline by 14 percent.

Services—Broad-based Decline across Army Services Contracting Portfolio in 2015

Figure 7-4: Army Services Contract Obligations by Category, 2000–2015

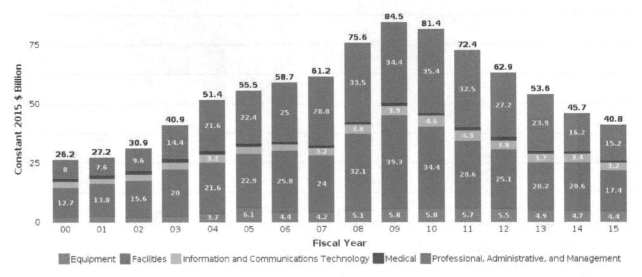

Source: FPDS; CSIS analysis.

Army contract obligations for services since 2012 have declined by 35 percent, roughly in line with the overall decline in Army contract obligations, and significantly steeper than the decline in overall DoD services contract obligations. Army services contract obligations declined by 15 percent in 2014, again roughly in parallel to the overall Army decline, but in 2015, Army services contract obligations declined by 11 percent, nearly double the rate of decline for overall Army, and slightly steeper than the rate of decline for overall DoD services contracts.

The main source of this decline in 2015 was a 15 percent drop in contract obligations for Facilities-related Services & Construction (FRS&C), a decline of $3.2 billion. Within FRS&C, there was a nearly $800 decline in obligations for "Construction of Other Non-Building Facilities," but otherwise the decline was broad-based across the category of services. Professional, Administrative, and Management Support (PAMS) services declined by 6 percent in 2015, with a nearly $750 million increase in obligations for "Education/Training – Training/Curriculum Development" outweighed by significant declines of $200–$400 million in obligations for "Education/Training – Other," "Logistics Support Services," and the unhelpfully labeled "Other Professional Services."

For the other services categories, Equipment-related Services (ERS) (-6 percent) and Information & Communications Technology (ICT) services (-5 percent) both declined roughly in parallel with overall Army services in 2015, while Medical (MED) services declined by 32 percent, though that represented a decline of only $260 million.

For analysis of recent trends in Army R&D contracting, see the Army subsection of the "R&D Contracting During the Budget Drawdown" section in Chapter 4.

Competition for Army Contract Obligations—Quality & Quantity of Competition for Army Contracts Increases Moderately during Budget Drawdown

Within the Army's contracting portfolio, the rate of effective competition increased from 45 percent in 2009 to 57 percent in 2013, before falling back to 52 percent by 2015. This fluctuation is heavily influenced by the overall changes in the Army's contracting portfolio since 2008: the declining share obligated for products, as well as the dearth of obligations for major weapons systems, were likely the major drivers of the increase in competition rate. There has also been a shift in the quality of competition for Army contract obligations since 2008, as the share of effectively competed Army contract obligations receiving 5 or more offers has risen from 30 percent in 2008 to 41 percent in 2015. Over that same period, the share of overall Army contract obligations that were competitively sourced but received only one offer fell by more than half, from 15 percent in 2008 to 7 percent in 2015; unlike for DoD overall, that decline has not stagnate in the last few years.

Looking at Army contract obligations by area, there has been no distinct trend in effective competition rates for products or services. For Army products contracts, roughly 27 percent were obligated after effective competition between 2008 and 2015, which is below the average rate for overall DoD products contracts over that period. Just over 70 percent of Army services contract obligations have received effective competition from 2008-2015, which is slightly above the rate observed for DoD services overall.

For Army R&D, the rate of effective competition rose from 38 percent in 2009 to a high of 53 percent in 2013, before falling back to 42 percent by 2015. The rise in effective competition rate is primarily the result of the disappearance of SD&D contract obligations related to the canceled Future Combat Systems program. The decline in effective competition since 2013 can be attributed to increasing sole-source contract obligations for ACD&P; as the share of Army R&D contract obligations going for ACD&P has increased since 2013, the rate of effective competition for those contract obligations has fallen from 59 percent in 2013 to just 13 percent in 2015. Army R&D contract obligations also frequently receive only one offer to competitively sourced solicitations—over 20 percent of Army R&D contract obligations were competitively sourced, but received only one offer, in all but one year since 2010, which notably exceeds the single-offer competition rate for DoD R&D overall.

Table 5: Top 20 Army Vendors, 2009 and 2015

Top 20 Contractors in 2009	Obligations in 2015 Millions	2008 Rank	Top 20 Contractors in 2015	Obligations in 2015 Millions	2014 Rank
General Dynamics	7,437	2	Lockheed Martin	3,917	4
Oshkosh	6,503	18	Raytheon	3,640	3
Raytheon	6,197	5	General Dynamics	2,870	2
Boeing	5,801	4	United Technologies	2,150	5
KBR	5,132	3	Northrop Grumman	1,890	6
Top 5 Total	**31,069**		**Top 5 Total**	**14,467**	
Lockheed Martin	4,911	9	Boeing	1,855	1
BAE Systems	4,086	1	BAE Systems	1,469	7
Northrop Grumman	2,979	8	L3 Communications	1,319	8
AM General	2,935	7	Oshkosh	1,180	23
United Technologies	2,753	6	SAIC	916	9
L3 Communications	2,361	10	AM General	852	61
ITT	2,107	11	Vectrus	734	N/A
SAIC	1,982	14	ATK	721	12
URS	1,647	16	Fluor	706	11
ATK	1,600	13	General Atomics	653	19
Finmeccanica	1,452	20	CACI	627	13
Computer Sciences Corp.	1,315	15	Dyncorp International	617	17
Hensel Phelps	1,239	22	Booz Allen Hamilton	552	15
CACI	1,135	23	Airbus Group	548	35
Honeywell	1,090	21	Harris	527	16
Top 20 Total	**64,662**		**Top 20 Total**	**27,743**	
Overall Army Total	**162,086**		**Overall Army Total**	**72,293**	

Source: FPDS; CSIS analysis.

Three of the top 5 Army vendors in 2009 are not among the top 5 in 2015: Oshkosh, which was 9th in 2015 (up from 23rd in 2014), Boeing, which is 6th in 2015 (and which was 1st in 2014), and KBR, which is outside the top 20 in 2015. Replacing them in the top 5 are Lockheed Martin, which was ranked 6th in 2009, United Technologies, which ranked 10th in 2009, and Northrop Grumman, which was ranked 8th in 2009. Six vendors that were in the top 20 in 2009 are not in the top 20 in 2015: the aforementioned KBR, ITT, URS Finmeccanica, Computer Sciences Corporation, Hensel Phelps, and Honeywell. Meanwhile, four vendors in the top 20 in 2015 were not in the top 20 in 2014: the aforementioned Oshkosh, AM General (61st in 2014), Vectrus (a newly formed vendor), and EADS North America (35th in 2014.)

In 2009, the threshold to be ranked among the top 20 Army vendors was approximately $1.1 billion; by 2015, that threshold had fallen by more than half, to slightly over $500 million. There was no significant change in the degree of concentration within the Army's supporting industrial base between 2009 and 2015: top 5 share of overall Army contract obligations rose from 19 percent to 20 percent, while the top 20 share fell from 40 percent to 38 percent.

7.2. Navy—Year-to-Year Trends Driven by Timing of Large Production Contracts

Navy contract obligations have been strongly preserved since the initial impact of sequestration in 2012, falling by only 14 percent, slightly more than half the rate of decline for overall DoD contract obligations. 2014 saw an 11 percent decline, which was actually slightly steeper than the overall DoD decline in that year, but that was an artifact of the timing of contracts for F-35s—there was a large buy in FY2013, and the "decline" in Navy contract obligations in 2014 was primarily a reflection of a drop from that spike. In 2015, Navy contract obligations were virtually stable (-1 percent), as overall DoD contract obligations declined moderately. The data indicates that, putting aside the ebbs and flows of F-35 orders, Navy contract obligations have weathered the recent downturn far better than the other major DoD components. Figure 7-5 shows Navy contract obligations, broken down by area, from 2000–2015.

Figure 7-5: Navy Contract Obligations by Area, 2000–2015

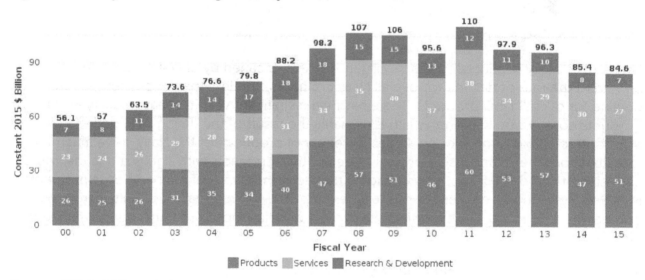

Source: FPDS; CSIS analysis.

Figure 7-6: Navy Products Contract Obligations by Category, 2000–2015

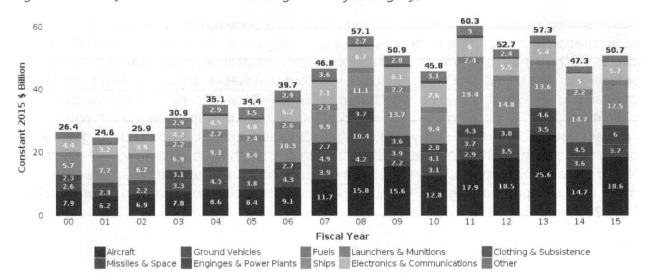

Source: FPDS; CSIS analysis.

Since 2012, Navy products contract obligations have declined by only 4 percent, less than one-third the rate of decline for overall Navy contracts, and less than one-sixth the rate of decline for overall DoD products. There has been significant year-to-year volatility in Navy products contract obligations, which is primarily the result of the timing of large F-35 block buy contracts. Ships, the second-largest category of products for the Navy, have a particular stability about them, because many ships are multiyear projects with significant lead-time, so cancelation and delay is more burdensome than for other types of major defense programs. Multiyear agreements also play a role, as the costs inherent to breaking a multiyear purchasing agreement serve a disincentive to cut or delay those programs during a budget drawdown.

In 2015, Navy Aircraft contract obligations rose by 26 percent, with a nearly $6 billion increase in obligations related to the F-35 program and a $900 million increase in obligations for the E-2C Hawkeye aircraft program, offsetting a nearly $2 billion decline in obligations related to the F/A-18 E/F program and a $1 billion decline in obligations related to the P-8A Poseidon aircraft program. Contract obligations for Engines & Power Plants grew by 34 percent in 2015, driven by a nearly $1.5 billion increase related to the F-35 program. Navy contract obligations for Electronics & Communications products rose by 14 percent in 2015, with increases of $200–$400 million in obligations for "Electronic Countermeasures & Quick Reaction Equipment," "Miscellaneous Communications Equipment," and "Radar Equipment, Except Airborne."

Navy contract obligations for Ships declined by 15 percent in 2015, after an 8 percent increase in 2014, with a $1 billion decline in obligations related to Nimitz-class aircraft carriers and a nearly $2 billion decline in obligations for Virginia-class submarines (after a one-year spike in 2014) offsetting increases of $200–$400 million for unlabeled "Combat Ships and Landing Vessels," "Fire Control Sonar Equipment," and the catch-all category "Miscellaneous Vessels."

For the other Navy products contract categories with significant levels of obligations, contract obligations for Launchers & Munitions (-11 percent) and Other products (-10 percent) declined significantly, while Missiles & Space (5 percent) saw a modest increase.

Services—Decline in Navy Services Contract Obligations Accelerates in 2015, Driven by Reductions in IT-related Contracts

Figure 7-7: Navy Services Contract Obligations by Category, 2000–2015

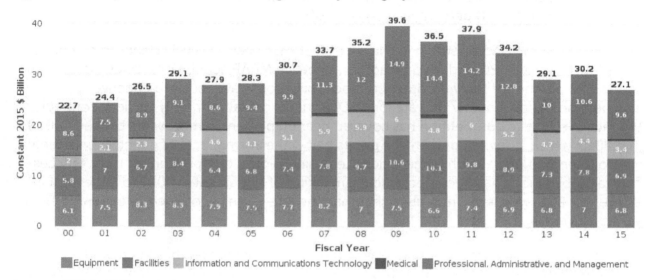

Source: FPDS; CSIS analysis.

Navy contract obligations for services declined by 21 percent between 2012 and 2015, significantly more steeply than the overall decline for Navy contracts, but roughly in line with the overall decline in DoD services contracts. Navy services contract obligations increased by 4 percent in 2014, but declined by 10 percent in 2015, in line with acceleration of the decline in services contracts within DoD as a whole.

The largest source of decline within the Navy's services contract portfolio in 2015 was for ICT, which declined by 21 percent, driven by a $350 million decline for "IT & Telecom – Telecommunications Network Management" and a nearly $800 million decline for the unhelpfully vague "IT & Telecom – Other IT & Telecommunications."

The other four categories of services within the Navy all saw declines of varying magnitudes in 2015: ERS (-4 percent) and MED (-2 percent) were relatively preserved, while FRS&C (-11 percent) and PAMS (-10 percent) declined roughly in parallel to the overall decline in Navy services contract obligations.

R&D

For analysis of recent trends in Navy R&D contracting, see the Navy subsection of the "R&D Contracting During the Budget Drawdown" section in Chapter 4.

Since 2008, the rate of effective competition for Navy contract obligations has fallen from 43 percent to 35 percent. The decline has primarily come in competitions receiving 5 or more offers: the share of overall Navy contract obligations that were competitively sourced and received 5 or more offers fell from 18 percent in 2008 to 11 percent in 2015. Single-offer competition declined from 12 percent of Navy contract obligations in 2010 to 7 percent in 2013, but rose back to 8 percent in 2014 and 2015.

The rate of effective competition for Navy products has been relatively steady in recent years, aside from a spike in 2008 due to broadly competitive MRAP acquisitions: since 2009, roughly 20 percent of Navy products contract obligations have received effective competition, notably below the rate for DoD products overall. For Navy services contracts, the rate of effective competition has hovered near 60 percent in most years between 2008 and 2015, which is several percentage points below the average rate of effective competition for DoD services overall. While the incidence of single-offer competition for Navy services contract obligations has been declining (from a high of 17 percent in 2010 to 11 percent in 2015), that 2015 rate is still notably higher than the rate of single-offer competition for DoD services overall.

The rate of effective competition for Navy R&D contract obligations has fluctuated significantly throughout the 2008-2015 period, reaching a low of 44 percent in 2008 and a high of 58 percent in 2013; in 2015, 49 percent were effectively competed, which is roughly in line with the rate for DoD R&D overall.

Table 6: Top 20 Navy Vendors, 2009 and 2015

Top 20 Contractors in 2009	Obligations in 2015 Millions	2008 Rank	Top 20 Contractors in 2015	Obligations in 2015 Millions	2014 Rank
Lockheed Martin	13,787	1	Lockheed Martin	15,510	1
Northrop Grumman	9,045	2	General Dynamics	7,645	2
General Dynamics	8,950	3	Raytheon	3,582	4
Raytheon	5,816	5	Boeing	3,554	3
Boeing	5,573	4	United Technologies	3,078	9
Top 5 Total	**43,170**		**Top 5 Total**	**33,369**	
Bell-Boeing Joint Project Office*	2,944	8	Northrop Grumman	3,006	7
BAE Systems	2,535	6	Huntington Ingalls	2,798	5
United Technologies	2,407	9	BAE Systems	2,497	6
Bechtel	2,121	11	Bechtel	2,485	8
SAIC	1,715	12	Bell-Boeing Joint Project Office*	2,029	10
L3 Communications	1,558	14	Textron	994	14
General Electric	1,427	15	Austal	842	16
Hewlett-Packard	1,283	162	SAIC	822	12
ITT	985	16	Hewlett-Packard	818	11
Navistar	929	7	General Electric	759	15
Textron	723	17	Department of Energy	700	21
Computer Sciences Corp.	690	19	L3 Communications	591	13
Force Protection	571	13	Johns Hopkins APL	548	18
Booz Allen Hamilton	546	26	Booz Allen Hamilton	482	17
Johns Hopkins APL	544	23	Charles Stark Draper Laboratory	468	20
Top 20 Total	**64,149**		**Top 20 Total**	**53,208**	
Overall Navy Total	**105,603**		**Overall Navy Total**	**84,576**	

Source: FPDS; CSIS analysis. (* - Joint Venture).

The only change to the top 5 Navy vendors is the result of Huntington Ingalls Industries spin-off from Northrop Grumman: United Technologies, which ranked 8th in 2009, moved into the top 5 in 2015, while Northrop Grumman and Huntington Ingalls ranked 6th and 7th, respectively, in 2015. Four vendors that were among the top 20 Navy vendors in 2009 were not among the top 20 in 2015: ITT, Navistar, Computer Sciences Corporation, and Force Protection (an MRAP producer acquired by General Dynamics in 2011.) And only one vendor in the top 20 in 2015 was outside the top 20 in 2014: the U.S. Department of Energy, which ranked 21st in 2014.

In 2009, the threshold to be ranked among the top 20 Navy vendors was roughly $540 million; in 2015, that threshold (roughly $470 million) had declined far less for the Navy than it had for the other major DoD components, or for any of the three contracting market areas. As with the Army, the level of concentration within the Navy's supporting industrial base was relatively unchanged during the budget drawdown: the top 5 share of overall Navy contract obligations declined from 41 percent to 39 percent between 2009 and 2015, while the top 20 share rose from 61 percent to 63 percent.

7.3. Air Force—Air Force Decline Broadly Parallels Decline for DoD Overall

Since 2012, Air Force contract obligations have declined by 27 percent, roughly in parallel to the overall decline in DoD contract obligations in the 2012–2015 period. Contract obligations within the Air Force were virtually stable (0 percent) in 2014, but declined by 7 percent in 2015, which was only slightly steeper than the decline in overall DoD contract obligations. Figure 7-8 shows Air Force contract obligations, broken down by area, from 2000–2015:

Figure 7-8: Air Force Contract Obligations by Area, 2000–2015

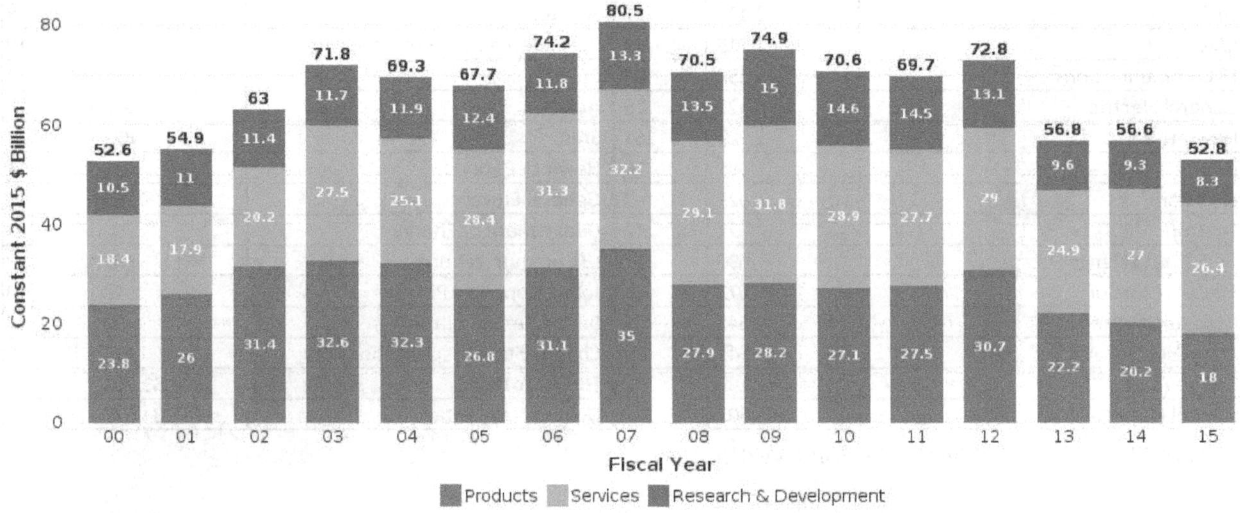

Source: FPDS; CSIS analysis.

Figure 7-9: Air Force Products Contract Obligations by Category, 2000–2015

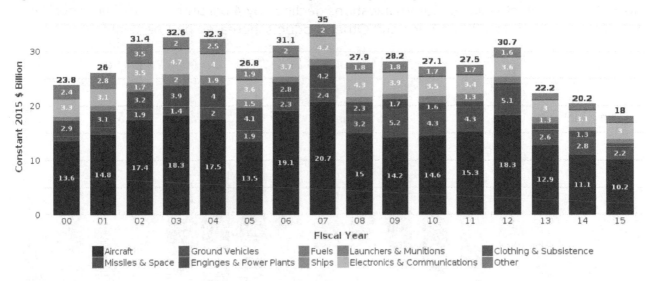

Source: FPDS; CSIS analysis.

Since 2012, Air Force products contract obligations have declined by 41 percent, more than half again as steeply as both overall Air Force contract obligations and overall DoD products contract obligations. Contract obligations for Air Force products declined by 9 percent in 2014, and declined by another 11 percent in 2015.

The main sources of decline within the Air Force's products contract portfolio in 2015 were Aircraft (-9 percent), Engines & Power Plants (-59 percent), and Missiles & Space (-21 percent). For Aircraft, the drivers of this decline were a nearly $700 million decline in obligations related to the C-5 RERP engine upgrade program, an over-$900 million decline in obligations related to the KC-45A tanker,[106] and declines of $200–$400 million related to the C-130J transport aircraft, the C-17A transport aircraft, and the T-6A training aircraft. These declines in 2015 outweighed a $1.1 billion increase in obligations for "Aircraft, Fixed Wing" that are not labeled by program in FPDS, as well as a nearly $500 million increase for "Miscellaneous Aircraft Accessories and Components."

For Engines & Power Plants, the decline in 2015 was driven by a $700 million drop in obligations for "Gas Turbines & Jet Engines, Aircraft" not labeled by program in FPDS. Within the Air Force's Missiles & Space contract portfolio, contract obligations related to the JASSM cruise missile program declined by nearly $500 million (after a one-year spike in 2014), obligations related to the NAVSTAR GPS satellite program fell by $600 million, and obligations for the SBIRS HIGH satellite program declined by over $400 million. These

---

[106] In past reports, CSIS noted over a billion dollars a year of Air Force products contract obligations were labeled as going to the Shillelagh Missile, a 1970s Army antitank missile program, due to the Air Force's reusing of System Equipment Codes. By investigating the individual contracts, CSIS discovered that those contracts were actually for the KC-45A tanker program, and in 2015, the Air Force properly labeled all KC-45A contract obligations.

declines outweighed a nearly $900 million increase related to the AIM-120 AMRAAM air-to-air missile program.

For the other Air Force products categories with significant levels of obligations, contract obligations for Electronics & Communications declined by 4 percent in 2015, Launchers & Munitions increased by 24 percent, and Other products increased by 1 percent.

Services—Air Force Services Contract Obligations Broadly Preserved during Budget Drawdown

Figure 7-10: Air Force Services Contract Obligations by Category, 2000–2015

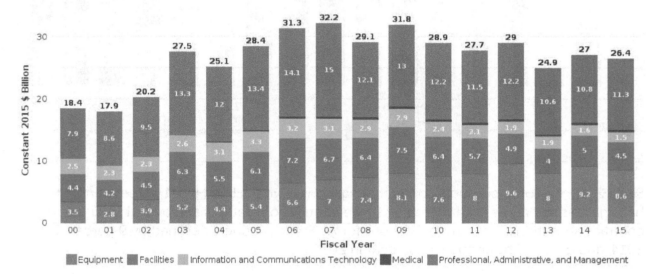

Source: FPDS; CSIS analysis.

Air Force services contract obligations have been strongly preserved since 2012, declining by only 9 percent, one-third the rate of decline for overall Air Force contracts, and significantly lower than the rate of decline for overall DoD services contracts. Furthermore, contract obligations for Air Force services increased by 8 percent in 2014, and declined slightly (-2 percent) in 2015.

Despite the relative stability of overall Air Force services contract obligations in 2015, there were significant disparities in trends between the different categories of services. FRS&C (-10 percent) declined at five times the rate of overall Air Force services, and ICT (-6 percent) declined at three times the rate of overall Air Force services. ERS (-6 percent) also declined at three times the rate of overall Air Force services, driven by a $700 million decline related to the EELV space launch program (after a spike in 2014), a $300 million fall related to Maintenance/Repair for the KC-10A transport aircraft, and a $200 million decline for Maintenance/Repair of Electrical Equipment related to the E-8A aircraft program. The category also saw notable increases in 2015: a $300 million increase in obligations for Maintenance/Repair of Aircraft Components related to the F-15 fighter program, a nearly $200 million increase related to Maintenance/Repair of Engines & Turbines for the F-22 fighter program, and a $300 million increase in Maintenance/Repair of Aircraft not labeled by program in FPDS.

Air Force contract obligations for PAMS increased by 5 percent in 2015, driven by a $1.1 billion increase in obligations for Engineering & Technical Services and a nearly $700 million increase in obligations for the catch-all category Other Professional Services, offsetting a $400 million decline related to the C-17A transport aircraft program. Meanwhile, MED increased by 10 percent in 2015, though that represented an increase of only approximately $40 million.

R&D

For analysis of recent trends in Air Force R&D contracting, see the Air Force subsection of the "R&D Contracting During the Budget Drawdown" section in Chapter 4.

Competition for Air Force Contract Obligations—Rate of Competition for Air Force Services Contracts Continues to Plummet

With the exception of a brief spike in 2013 and 2014, the rate of effective competition for Air Force products contract obligations has remained between 17 percent and 20 percent since 2008. For Air Force R&D, the rate of effective competition fell from 53 percent in 2008 to 37 percent in 2011, but has risen back to 48 percent in 2014 and 2015.

In an October 2015 report on defense competition, CSIS noted that the Air Force has seen a significant decline in its level of competition for services contract obligations, in a period where competition rates for services in other parts of DoD are either stable or rising.[107] While about one-fourth to one-third of that decline was explainable by shifts in the mix of services that the Air Force was contracting for and the reclassification of space launches (which had been almost entirely sole-source) from a product to a service, the data nonetheless showed a real and significant decline in the rate of effective competition for services within the Air Force.

Figure 7-11 shows the rates of effective competition for overall Air Force services, and for the five different categories of government services contracts—Equipment-related Services (ERS), Facilities-related Services & Construction (FRS&C), Information & Communication Technology (ICT) services, Medical (MED) services, and Professional, Administrative, and Management Support (PAMS) services—since 2010, along with the 2015 rates of effective competition for those categories outside the Air Force (the black striped bar at the bottom of each group) for comparison:

[107] Jesse Ellman, "Air Force Faces Puzzling Decline in Competition for Services," Center for Strategic & International Studies, October 2015, https://csis-prod.s3.amazonaws.com/s3fs-public/legacy_files/files/publication/150925_Ellman_AirForceFacesPuzzlingDecline.pdf.

Figure 7-11: Rate of Effective Competition for Air Force Services Contract Obligations, by Category of Services, 2010–2015

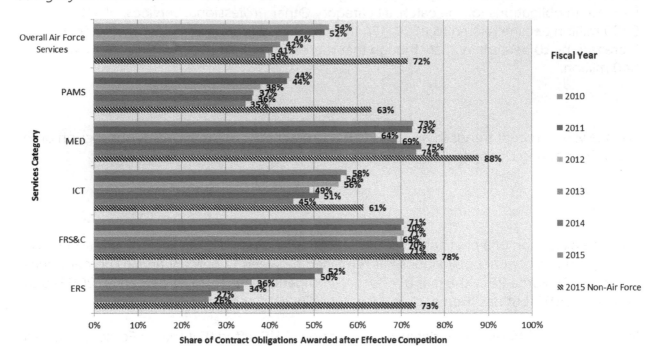

Source: FPDS; CSIS analysis.

For overall Air Force services, the rate of effective competition (54 percent) was already notably lower than the rate for non-Air Force services (70 percent) in 2010. That gap has only widened in recent years: while the rate of effective competition for non-Air Force services has risen slightly, to 72 percent in 2015, the rate for Air Force services has fallen to 39 percent in 2015, only slightly more than half the rate for non-Air Force services.

This decline in effective competition for services within the Air Force, both in absolute terms and relative to non-Air Force services contracts, is broad-based within the Air Force's services contracting portfolio, spanning three of the five categories of services, accounting for 80 percent of Air Force services contract obligations since 2010:

- For ERS, which has accounted for 31 percent of Air Force services contract obligations since 2010, the rate of effective competition fell by half, from 52 percent in 2010 to 26 percent in 2015, even as the rate for non-Air Force ERS rose from 60 percent to 73 percent.

- For PAMS, which has accounted for 42 percent of Air Force services contract obligations since 2010, the rate of effective competition has fallen from 44 percent in 2010 to 35 percent in 2015, as the rate for non-Air Force PAMS hovered near two-thirds.

- For ICT, which has accounted for 7 percent of Air Force services contract obligations since 2010, the rate of effective competition fell from 58 percent in 2010 to 45 percent in 2015; meanwhile, the rate for non-Air Force ICT was 61 percent in 2015, and fluctuated between 52 percent and 61 percent during the 2010 to 2015 period.

In the report linked above, CSIS also noted that this decline does not appear to be the result of the Air Force buying different specific types of services (as defined by government Product or Service Codes (PSCs)) within those service categories that traditionally receive less effective competition than the rest of DoD; rather, the data shows that the Air Force has gotten lower, and in many cases declining, rates of competition than non-Air Force DoD for the same specific types of services.

Table 7 shows the rates of effective competition for Air Force, and non-Air Force DoD, for six specific types of services under ERS, which have collectively accounted for over three-quarters of Air Force ERS contract obligations since 2010:

Table 7: Effective Competition for PSCs under ERS, Air Force vs. non-Air Force DoD

Type of Equipment-related Service	2010 Air Force Effective Competition Rate	2015 Air Force Effective Competition Rate	2015 Non-Air Force Effective Competiton Rate
Maintenance/Repair of Aircraft	63%	52%	85%
Maintenance/Repair of Aircraft Components	9%	4%	76%
Maintenance/Repair of Electrical Equipment	1%	3%	67%
Maintenance/Repair of Engines & Turbines	48%	5%	10%
Maintenance/Repair of Miscellaneous Equipment	80%	78%	46%
Maintenance/Repair of Weapons	93%	32%	1%

Source: FPDS; CSIS analysis.

For both Maintenance/Repair of Aircraft Components and Maintenance/Repair of Electronics Equipment, the rates of effective competition within the Air Force have been low throughout the period, while the rates for non-Air Force are not just higher, but notably high in absolute terms. For Maintenance/Repair of Aircraft and Maintenance/Repair of Weapons, Air Force has seen significant declines in effective competition since 2010, although for the latter, that rate still far exceeds the rate for non-Air Force. The rate of effective competition for Maintenance/Repair of Engines & Turbines was actually relatively high until 2015, when it cratered to below non-Air Force levels. And for Maintenance/Repair of Miscellaneous Equipment, the Air Force actually gets significantly higher rates of effective competition.

Table 8 shows the rates of effective competition for Air Force, and non-Air Force DoD, for six specific types of services under PAMS, which have collectively accounted for approximately 70 percent of Air Force PAMS contract obligations since 2010.

Table 8: Effective Competition for PSCs under PAMS, Air Force vs. non-Air Force DoD

Type of Professional, Administrative, and Management Support Service	2010 Air Force Effective Competition Rate	2015 Air Force Effective Competition Rate	2015 Non-Air Force Effective Competiton Rate
Engineering & Technical Services	52%	31%	63%
Logistics Support Services	8%	17%	75%
Other Professional Services	35%	19%	67%
Program Management/Support Services	59%	52%	74%
Systems Engineering Services	21%	36%	102%
Technical Representative Services - Guided Missiles	93%	75%	-9%

Source: FPDS; CSIS analysis.

For Engineering & Technical Services, Other Professional Services, and Program Management Support Services, and Technical Representative Services – Guided Missiles, the rates of effective competition within the Air Force have been declining since 2010, and for the first three, those rates were already lower than the rates for non-Air Force. The rates of effective competition for Logistics Support Services and Systems Engineering Services have actually been rising within the Air Force, but are still well below the rate for non-Air Force.

Table 9 shows the rates of effective competition for Air Force, and non-Air Force DoD, for six specific types of services under ICT, which have collectively accounted for over 75 percent of Air Force ICT contract obligations since 2010.

Table 9: Effective Competition for PSCs under ICT, Air Force vs. non-Air Force DoD

Type of Information & Communications Technology Service	2010 Air Force Effective Competition Rate	2015 Air Force Effective Competition Rate	2015 Non-Air Force Effective Competiton Rate
Automated Information System Services	53%	66%	71%
IT & Telecom - Other IT & Communications	62%	39%	64%
IT & Telecom – Programming	43%	69%	55%
IT & Telecom – Telecommunications Network Management	79%	51%	80%
Maintenance/Repair of ADP Equipment & Supplies	93%	23%	45%
Maintenance/Repair of Communications Equipment	17%	4%	46%

Source: FPDS; CSIS analysis.

For four of the six PSCs (IT & Telecom – Other IT & Communications, IT & Telecom – Telecommunications Network Management, Maintenance/Repair of ADP Equipment & Supplies, and Maintenance/Repair of Communications Equipment), the rate of effective competition within the Air Force has fallen significantly between 2010 and 2015, to levels significantly below the rate of effective competition for non-Air Force. For Automated Information System Services, the rate of effective competition within the Air Force has increased notably, but is still below the rate for non-Air Force. Whereas for IT & Telecom – Programming, the rate of effective competition has increased to a level above that for non-Air Force.

This data, taken collectively, shows that there has been a real and significant decline in the rate of effective competition for services contracts within the Air Force. Not only is the Air Force getting lower and declining rates of effective competition for the same categories of services than non-Air Force DoD, it is often getting lower and declining rates of effective competition for the same specific types of services.

Top 20 Air Force Vendors—Top Vendors for Air Force Contracts Unchanged during Budget Drawdown

Table 10: Top 20 Air Force Vendors, 2009 and 2015

Top 20 Contractors in 2009	Obligations in 2015 Millions	2008 Rank	Top 20 Contractors in 2015	Obligations in 2015 Millions	2014 Rank
Lockheed Martin	13,604	1	Lockheed Martin	7,389	1
Boeing	9,437	2	Boeing	7,271	2
Northrop Grumman	6,402	3	Northrop Grumman	3,546	3
Raytheon	4,025	4	Raytheon	2,897	6
L3 Communications	2,844	5	L3 Communications	2,648	4
Top 5 Total	**36,313**		**Top 5 Total**	**23,752**	
MIT	1,910	15	United Launch Alliance*	1,730	5
United Technologies	1,670	6	General Atomics	1,298	10
General Dynamics	1,197	7	United Technologies	1,010	7
Booz Allen Hamilton	1,058	9	MIT	934	8
Computer Sciences Corp.	1,050	8	Aerospace Corp.	840	13
Aerospace Corp.	875	13	Sierra Nevada	705	14
ITT	864	12	BAE Systems	582	12
Rockwell Collins	853	14	Alion Science & Technology	450	15
BAE Systems	775	10	Booz Allen Hamilton	398	11
General Atomics	718	17	Gilbane Building Company	335	54
SAIC	644	18	Wyle Laboratories	333	18
URS	537	16	Dyncorp International	330	17
General Electric	498	11	URS	323	24
Kelly Aviation Center	484	19	General Dynamics	320	16
Sierra Nevada	470	28	General Electric	309	9
Top 20 Total	**49,916**		**Top 20 Total**	**33,649**	
Overall Air Force Total	**74,926**		**Overall Air Force Total**	**52,743**	

Source: FPDS; CSIS analysis. (Joint venture).*

Unlike the other major DoD components or contracting market areas, the top 5 Air Force vendors were completely unchanged, down to the precise ranking, between 2009 and 2015. Despite that stability at the top level, there were five vendors among the top 20 Air Force vendors in 2009 who were not in the top 20 in 2015: Computer Sciences Corporation, ITT, Rockwell Collins, SAIC, and Kelly Aviation Center (which was acquired by Lockheed Martin in 2013). Two vendors in the top 20 in 2015 were not among the top 20 Air Force vendors in 2014: Gilbane Building Company (54th in 2014) and URS (24th in 2014).

In 2009, the threshold to be among the top 20 Air Force vendors was $470 million, but by 2015, that threshold had fallen to roughly $310 million. There was a minor decline in the level of concentration within the Air Force's supporting industrial base since 2009: the top 5 share of overall Air Force contract obligations declined from 48 percent to 45 percent, while the top 20 share fell from 67 percent to 64 percent.

7.4. Defense Logistics Agency[108]—DLA Contract Obligations Decline Roughly in Sync with Overall DoD Decline

DLA contract obligations declined by 7 percent between 2013 and 2014, slightly less steeply than for DoD overall. In 2015, contract obligations within DLA fell by 8 percent, which was somewhat steeper than the decline for overall DoD contract obligations. Figure 7-12 shows DLA contract obligations, broken down by area, from 2000–2015.

Figure 7-12: DLA Contract Obligations by Area, 2000–2015

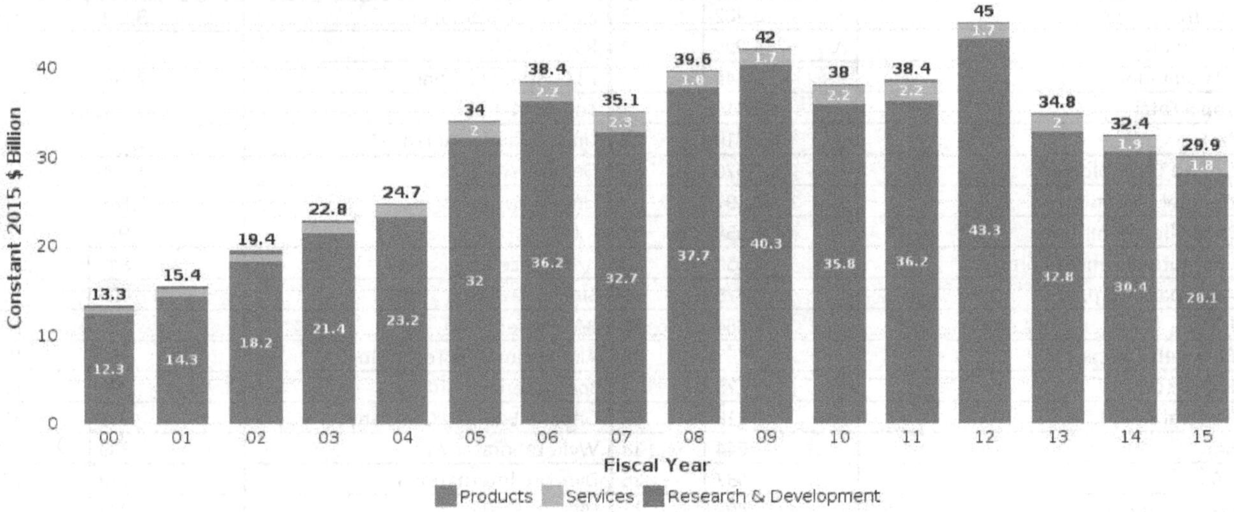

Source: FPDS; CSIS analysis.

[108] This section focuses on the 2013–2015 period for analysis, due to a one-year spike in DLA contract obligations for Fuels in 2012, related to the timing of large Fuels contracts, which distorts trend analysis.

Figure 7-13: DLA Products Contract Obligations by Category, 2000–2015

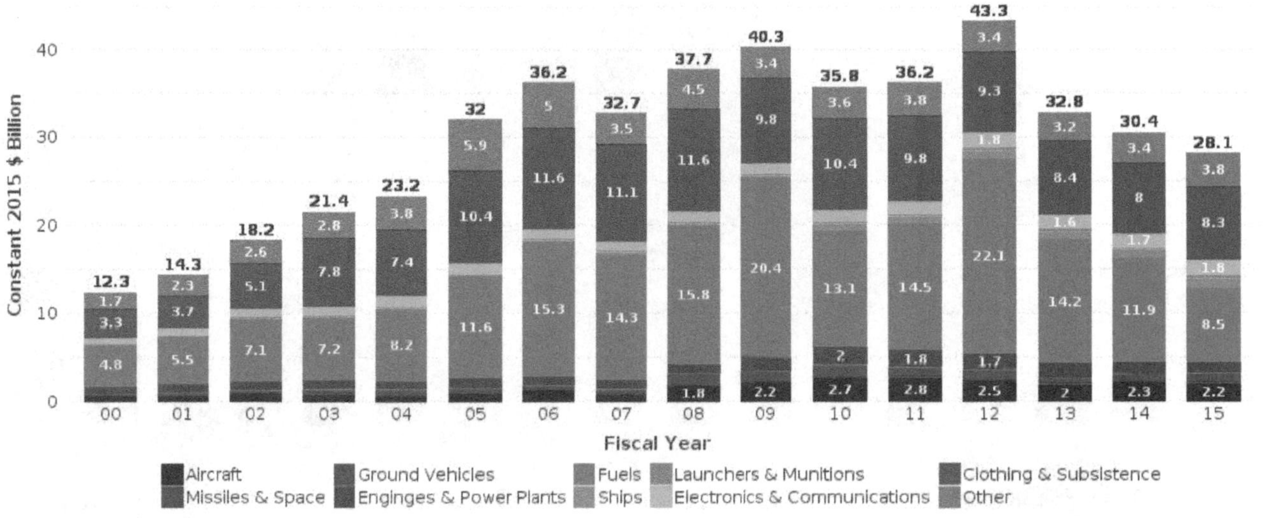

Source: FPDS; CSIS analysis

Source: FPDS; CSIS analysis.

Products, which have accounted for 94 percent of DLA contract obligations from 2013–2015, unsurprisingly declined at the same rates as overall DLA contract obligations in 2014 and 2015. The 8 percent decline in 2015, however, masks markedly varying trends within the different categories of products. The main source of decline within DLA's products contract portfolio was Fuels, the largest category of products within DLA; contract obligations declined by 28 percent in 2015, with a nearly $2.4 billion decline for Liquid Propellants – Petroleum Base and an $800 million decline for Fuel Oils. Engines & Power Plants (-10 percent) declined slightly more steeply than overall DLA products, while contract obligations related to Aircraft (-5 percent) were relatively preserved.

Every other category of products that had significant obligations within DLA in 2015 saw increases in obligations: Clothing & Subsistence (5 percent, driven by a $600 million increase for Drugs & Biologicals), Electronics & Communications (7 percent), Ground Vehicles (21 percent), Ships (25 percent), and Other products (9 percent).

Figure 7-14: DLA Services Contract Obligations by Category, 2000–2015

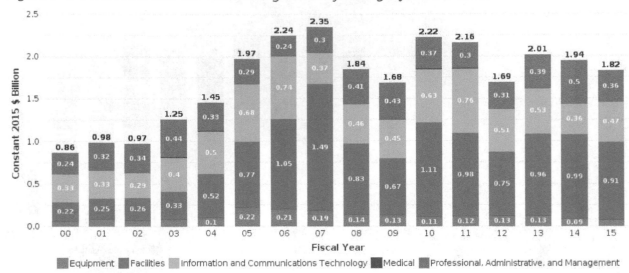

Source: FPDS; CSIS analysis.

DLA services contract obligations declined by 3 percent in 2014, which was less steep than the decline in overall DLA contract obligations for that year. In 2015, contract obligations for DLA services declined by 6 percent, which was slightly less steep than overall DLA contracts.

In 2015, DLA ICT contract obligations increased by 30 percent, while obligations for FRS&C (-7 percent) declined roughly in parallel with overall DLA services, and obligations for PAMS (-28 percent) declined at nearly five time the rate of DLA services contract obligations overall.

R&D

DLA does not obligate significant contract dollars for R&D.

7.5. Missile Defense Agency—MDA Sees Significant Declines in 2014 and 2015[109]

Since 2012 MDA contract obligations have declined by 34 percent, notably more steeply than for overall DoD contract obligations. Most of that decline was in 2014 and 2015: in 2014, MDA contract obligations declined by 22 percent, over twice the rate of DoD overall, and in 2015, MDA contract obligations fell by 24 percent, nearly five times the rate of DoD overall. Figure 7-15 shows MDA contract obligations, broken down by area, from 2000–2015.

---

[109] The data in this section does not reflect the massive reclassification of MDA R&D contract obligations that was discussed in Chapter 2.

Figure 7-15: MDA Contract Obligations by Area, 2000–2015

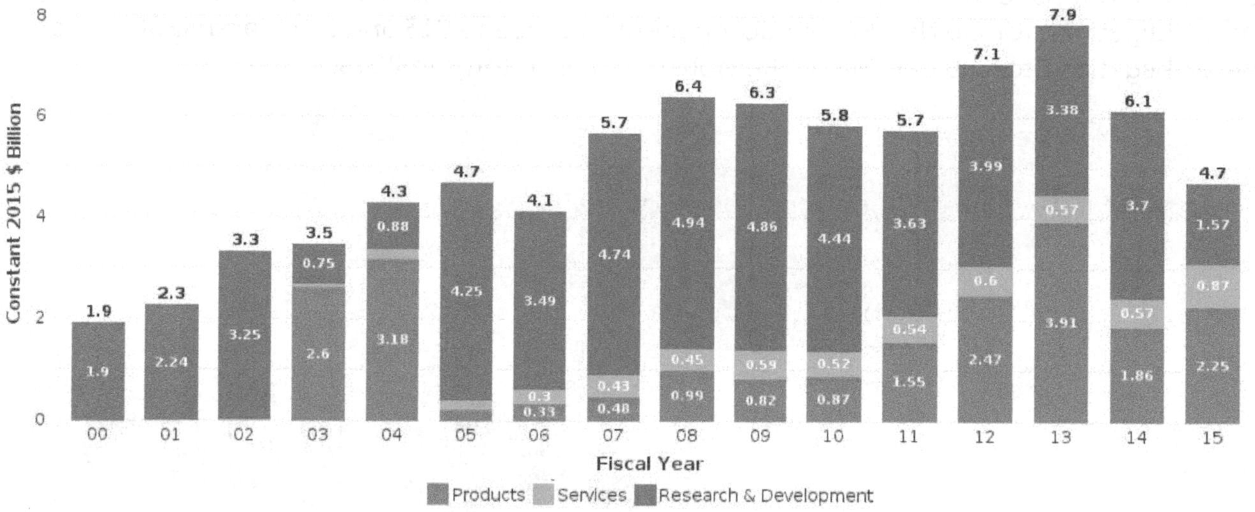

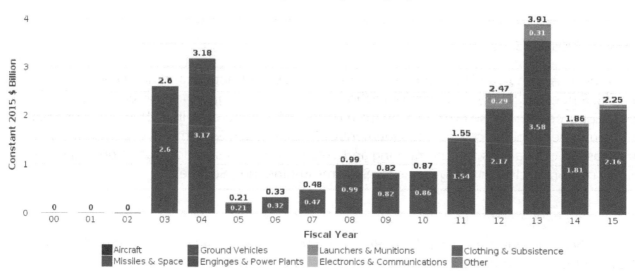

Source: FPDS; CSIS analysis.

Products—Significant Year-to-Year Volatility in MDA Products Contracts

Figure 7-16: MDA Products Contract Obligations by Category, 2000–2015

Source: FPDS; CSIS analysis.

MDA contract obligations for products have been highly volatile in recent years, reflecting the particular nature of MDA's products contract portfolio, which is centered on a few large, high-profile programs. For the 2012–2015 period, MDA products contract obligations have declined by 9 percent, roughly one-fourth the rate of MDA contract obligations overall, but that figure masks significant year-to-year fluctuations. MDA products contract obligations rose by 58 percent between 2012 and 2013, fell 52 percent in 2014, and rose by 21 percent in 2015, likely reflecting the timing of contracts for MDA's large programs.

Unsurprisingly, nearly all of MDA's contract obligations for products fall under the category of Missiles & Space, and the rates of change for that category of products closely track those of overall MDA products. Obligations for Guided Missiles increased by $1.5 billion between 2012 and 2013, but have fallen by $2.2 billion since 2013. Meanwhile, MDA saw $750 million of new obligations for Guided Missile Components between 2013 and 2015, and $200 million of new obligations for Guided Missile Remote Control Systems in 2015.

Services—Massive Increase in PAMS Obligations in 2015

Figure 7-17: MDA Services Contract Obligations by Category, 2000–2015

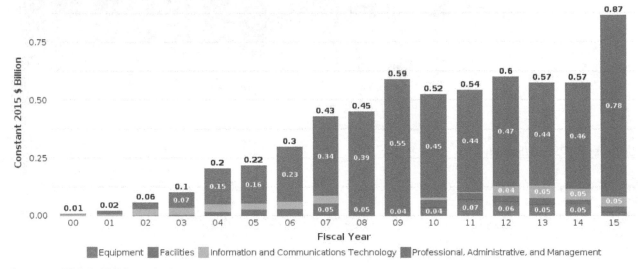

Source: FPDS; CSIS analysis.

MDA does less than a billion dollars a year in services contracting, and nearly all of that is for PAMS. Since 2012, MDA contract obligations for PAMS have grown by 66 percent, the result of a 71 percent increase in 2015. The main drivers of that increase in 2015 were over $300 million in new obligations for Management of Missile/Space Systems R&D and an $150 increase in obligations for Engineering and Technical Services, which outweighed a nearly $200 million decline in obligations for Systems Engineering Services.

R&D

For analysis of recent trends in MDA R&D contracting, see the MDA subsection of the "R&D Contracting During the Budget Drawdown" section in Chapter 4.

7.6. Other DoD—Contract Obligations Relatively Preserved Compared to Overall DoD

Since 2012, Other DoD contract obligations have declined by 17 percent, notably less steeply than for DoD contracts overall. Contract obligations for Other DoD were flat in 2014 (0 percent), and declined roughly in parallel (-6 percent) to the overall DoD decline in contract obligations in 2015. Figure 7-18 shows MDA contract obligations, broken down by area, from 2000–2015.

Figure 7-18: Other DoD Contract Obligations by Area, 2000–2015

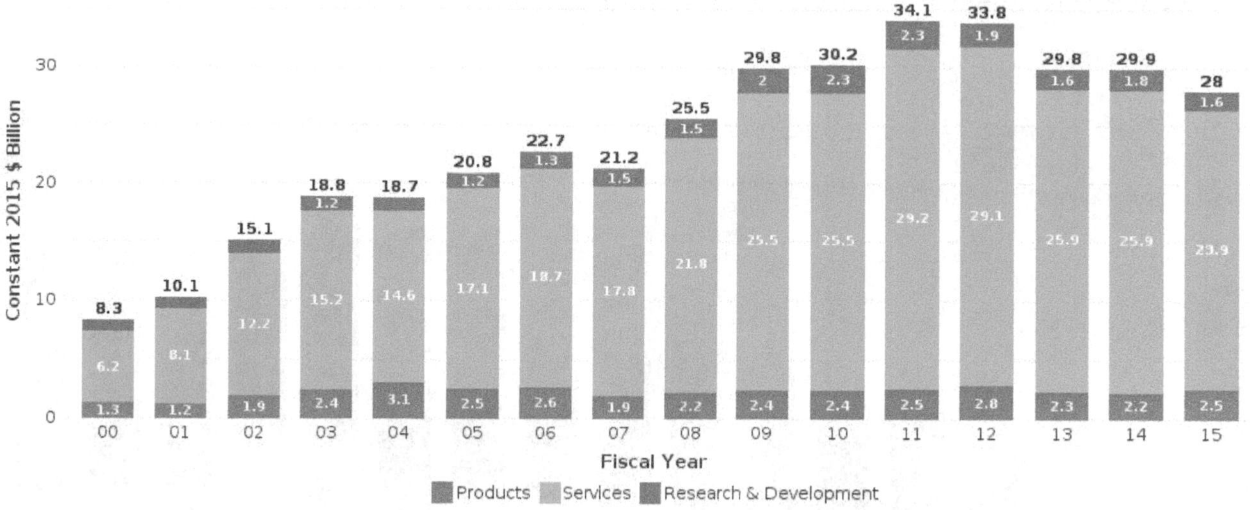

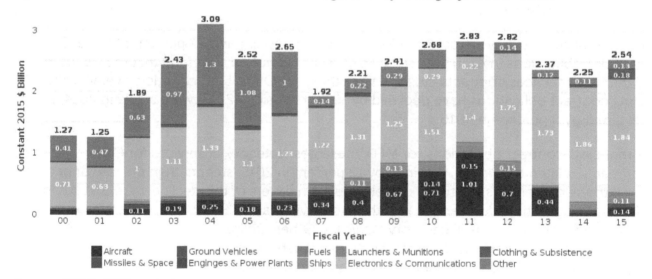

Source: FPDS; CSIS analysis.

Products—Significant, Broad-based Increase in Contract Obligations in 2015

Figure 7-19: Other DoD Products Contract Obligations by Category, 2000–2015

Source: FPDS; CSIS analysis.

Products make up a relatively small share of Other DoD contract obligations. Since 2012, Other DoD products contract obligations have been relatively preserved, declining by only 10 percent, less than two-thirds the rate of decline for Other DoD contract obligations overall, and two-fifths the rate of overall DoD products contracts. Other DoD products contract obligations decline by 4 percent in 2014, but grew by 12 percent in 2015.

The vast majority of Other DoD products contract obligations are for Electronics & Communications, which actually declined by 1 percent in 2015. The growth in Other DoD products contract obligations was entirely in categories with less than $200 million per year in obligations, most notably in Clothing & Subsistence, which saw a quadrupling of contract obligations between 2014 to 2015 (to roughly $180 million), driven by an $80 million increase in obligations for Medical & Surgical Instruments.

Services—Steep Declines in Obligations for Medical Services and Air Transport in 2015

Figure 7-20: Other DoD Services Contract Obligations by Category, 2000–2015

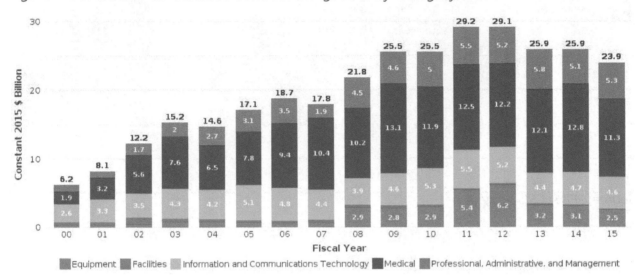

Source: FPDS; CSIS analysis.

Given that services have accounted for between 85 percent and 87 percent of Other DoD contract obligations in each year since 2012, it should not be surprising that the trends in Other DoD services contracts closely mirror those for Other DoD contracts overall. Other DoD contract obligations have declined by 18 percent since 2012, were stable in 2014, and declined by 8 percent in 2015.

Other DoD contract obligations for MED, the largest category of services for Other DoD (primarily related to Tricare), declined by 12 percent in 2015, steeper than the rate of decline for Other DoD services contracts overall. The main drivers of this decline were a $1 billon decline in obligations for General Health Care Services and a $400 million decline in obligations for the catch-all category Medical – Other. Obligations for ERS declined by 19 percent in 2015, over twice the rate of Other DoD services overall, driven by an over $300 million decline in obligations for Air Freight and an over $150 million decline in obligations for Maintenance/Repair of Aircraft, both related to US Transportation Command.

Other DoD contract obligations for ICT declined by only 3 percent in 2015, slightly more than one-third the rate of Other DoD services overall, driven by a $400 million decline in obligations for the unhelpfully vague IT & Telecom – Other IT & Telecommunications and a $200 million decline for Telecommunication Network Management, which outweighed a $300 million increase in obligations for Maintenance/Repair of Fiber Optics Materials. Other DoD contract obligations for PAMS, meanwhile, rose by 4 percent in 2015, with a $300 million decline in obligations for Passenger Air Charter Services outweighed by increases of between $100 and $200 million for Engineering and Technical Services, Logistics Support Services, and the catch-all category Other Professional Services.

For analysis of recent trends in Other DoD R&D contracting, see the Other DoD subsection of the "R&D Contracting During the Budget Drawdown" section in Chapter 4.

8. Conclusion

An Initial Look at FY2016 Defense Contract Trends

What does the FY2016 contract data say about the current state of defense acquisition?

With complete and reliable FY2016 DoD contract data only becoming available in early January 2017, CSIS has only begun to scratch the surface of what the data can tell about the current state of defense acquisition. Nonetheless, even an initial analysis has revealed several key findings. First, the data shows that the drawdown of DoD contract obligations is over; indeed, DoD contract obligations rose by 7 percent in 2016, after continuous decline since 2009. This increase was driven by increases in obligations for procurement of high-profile platforms, but both services and R&D showed minor gains as well.

The increase in obligations for services is noteworthy, because the acceleration in the decline of DoD services contract obligations in 2015 raised concerns that the long-feared "targeting" of services contracts for savings had begun. The 2016 data, however, shows that this acceleration was merely a one-year decline.

What Is DoD Buying?

What has happened with the Defense Innovation Initiative over the past year?

Over the past year, the Defense Innovation Initiative remained a top DoD priority. In the PB17 budget, DoD included $18 billion for Third Offset Strategy investments over the course of the FYDP. Rather than making large initial bets on a small set of capabilities, DoD elected to make numerous smaller bets. DIU(X) opened two new offices in Boston, Massachusetts, and Austin, Texas. Additionally, Secretary Carter created the Defense Innovation Board composed of a range of experts such as Amazon's Jeff Bezos, Code for America Founder Jennifer Pahlka, and scientist Neil deGrasse Tyson to identify a range of innovative private-sector practices and technological solutions which DoD can adapt.

Finally, in 2016 Secretary Carter declassified the Strategic Capabilities Office led by Dr. Will Roper. Originally established in 2012, the SCO works to ensure America's lead in military technological capabilities by mixing and merging technologies across multiple platforms and services. For example, SCO's work resulted in successful work on adapting the hypervelocity projectile for the Navy's existing conventional naval artillery gun pieces and making the Army's ATACMS capable of targeting naval targets. Perhaps most importantly, the SCO has achieved significant "buy in" from the military services as reflected by its increasing budgets each year.

What are the prospects for the Defense Innovation Initiative in the new administration?

At this time, the prospects for the Defense Innovation Initiative in the new administration remains unclear. Throughout the campaign, the president and his surrogates made little reference to military innovation focusing instead on "rebuilding the military." The selection of Retired Gen Mattis as secretary of defense provides little guidance given that he focused mostly on operational activities throughout his distinguished military career. However, the decision to retain, at least temporarily, Deputy Secretary Bob Work suggests that that military innovation might remain a DoD priority. Whomever is selected to eventually replace Deputy Secretary Work will be an indicator about the long-term future of the Defense Innovation Initiative.

Has the trough in DoD's development pipeline for major weapons systems persisted?

As many analysts and policy makers feared, DoD R&D contracts have borne a disproportionate share of cuts within the DoD contracting portfolio during the current budget drawdown. The dimensions of those cuts, however, have not followed the expected path. Despite fears that early-stage, seed-corn R&D would be hit particularly hard, the data show that it has been relatively preserved compared to the overall declines in R&D. In fact, within DoD, two categories of mid- to late-stage R&D, Advanced Technology Development (6.3) and System Development & Demonstration (6.5) have seen cuts of two-thirds or more between 2009 and 2015.

The two main drivers of the massive declines in those two stages of R&D are the cancelation of large R&D programs (such as the Army's Future Combat Systems) and the maturation of R&D programs into procurement (such as the F-35 Joint Strike Fighter). During the budget drawdown period, however, there has been a dearth of new development programs for major weapons systems that replace those that have either graduated into production or been canceled. In 2016, despite a slight increase in overall DoD R&D contract obligations, obligations for System Development & Demonstration (6.5) continued to decline sharply. As a result, DoD is facing what is now a seven-year trough in its development pipeline for major weapons systems.

How Is DoD Buying It?

What changes did the 2017 NDAA make in the defense acquisition system?

Between the House and Senate versions of the FY2017 NDAA, Congress enacted fundamental changes across three major elements of the defense acquisition system: how the system is organized and given its mission; how acquisition programs are structured; and what the business model is for defense research and development. The final conference agreement largely adopted House and Senate proposals to shift away from the traditional MDAP structure around which much of the defense acquisition system has centered, creating new openings for more prototyping, technology demonstrations, and acquisitions of commercial technology. This shift aligns with the decision to divide the responsibilities of the current USD AT&L between a USD R&E focused on early-stage technologies, and potentially on new business models for R&D, and a USD A&S focused on traditional MDAP programs. However, many of the statutory changes enacted will not be fully implemented until 2018, and the desire to reshape the defense acquisition system in Congress combined with the new administration's early interest in the topic suggests the likelihood for continued debate on these fundamental issues in the coming years. The incoming administration will need to quickly grapple with the changes in the FY2017 NDAA and determine how to balance the more incremental, internal DoD approach to acquisition improvement embodied in Better Buying Power with the more fundamental shifts desired by Congress, and incorporate these approaches into its own agenda for defense acquisition.

How has the Better Buying Power reform effort affected defense acquisition system performance?

There are many ways to measure and answer the question of how the performance of the defense acquisition system has responded to the last eight years of acquisition reform. Any answer to this question will be at best partial for years to come as we wait to learn the fate of major projects that were launched under these reforms, like the Ohio Replacement Submarine and the B-21 bomber. If these and smaller new projects fare well, the seven-year trough in the weapon system development pipeline could come to be viewed to some extent as a period of prudent consolidation taken to regroup and avoid costly failures. However, if these efforts suffer higher-than-average cost escalation or fail to deliver on their core technological promises, then complaints about stifled innovation during the trough period will be redoubled. At the same time, there is an increasing amount of data available on the programs and contracts that have already matured in this period. Multiple sources point in the same direction: the Better Buying Power reforms, perhaps in combination with congressional efforts like the Weapon System Acquisition Reform Act of 2009, appear to have made progress reducing unit cost growth, despite unfavorable circumstances. This finding relies not just on AT&L's internal studies, but also GAO reports, and DIIG's own work on contract outcomes.

Cost-focused measures in Better Buying Power, like investment in cost estimates or aligning industry's incentives with affordability, are not a cure-all. While cycle-time has not gotten worse, the quality of scheduling estimates degraded even as cost-estimates got better. Defense reform efforts are often fatalistically characterized as wheel spinning. That reading appears to be too fatalistic, and overlooks real progress that has been achieved that is worth preserving even should priorities shift toward new goals.

Has DoD usage of contract pricing types changed in response to pessimism from Congress?

In recent years, there has been significant pressure on DoD from Congress to increase the use of fixed-price contract types. During the budget drawdown, however, the usage levels of fixed price and cost reimbursement contract types in DoD contracting have been largely unchanged. This stability is broad-based across most of the major DoD components and across the range of what DoD contracts for, with usage within the products, services, and R&D contracting portfolios largely unchanged during the budget drawdown period.

There have, however, been significant shifts in the fee types used for both fixed price and cost reimbursement contracts within DoD. The importance of fee type has been shown in Under Secretary Kendall's 2014 Performance of the Defense Acquisition Report, which found no performance difference between fixed price and cost reimbursement, but significant benefits from use of incentive-fee contract types. Putting this finding into action, the share of fixed price contract obligations structured as Fixed Price Incentive Fee has risen seven-fold between 2008 and 2015, primarily within the DoD products and R&D contracting portfolios. Surprisingly, there has been no corresponding increase in the use of Cost Plus Incentive Fee contract types; the share of cost reimbursement contract obligations structured as Incentive Fee has fallen significantly since 2011, primarily within products contracts. Additionally, the share of cost reimbursement contract obligations structured as Cost Plus Award Fee has fallen by over three-fourths since 2007. In its place, use of Cost Plus Fixed Fee has risen dramatically, from roughly a third of cost reimbursement contract obligations in 2007 to over two-thirds in 2015.

From Whom Is DoD Buying?

How has the drawdown affected the composition of the defense industrial base?

Despite the massive decline in DoD contract obligations since 2009, the composition of the defense industrial base, as measured by size of vendor, has been relatively stable in the 2009–2015 period for Medium vendors, Large vendors, and the Big 5 vendors.

Small vendors, by contrast, have actually increased their share in the last two years, from 16 percent in 2009–2013 to 19 percent in 2014 and 2015; this increase is broad-based, across most of the major DoD components and within both the services and R&D contracting portfolios. Obligations to small vendors declined slightly in 2015, but nonetheless, Small is the only category of vendors for which 2015 contract obligations are higher than they were in 2013. This data can be seen as a victory for policies that promote Small business participation: despite the pressures of the budget drawdown, Small vendors have managed to not just maintain their place in the defense contracting marketplace, but increase it.

Of the three areas of the defense-contracting marketplace, R&D has seen by far the most dramatic shift in the composition of the supporting industrial base. In 2009, the Big 5 vendors accounted for 57 percent of DoD R&D contract obligations, but that share has declined steadily since, to just 33 percent in 2015, and declined further in 2016, to 29 percent. This is particularly notable because of the massive decline in DoD R&D contract obligations since 2009; overall, the Big 5 control roughly two-fifths less of a market that is less than half the size it was in 2009. The primary driver of this decline is the trough in DoD's development pipeline for major weapons systems. With a number of major development programs either maturing into production or getting canceled, and a dearth of new large development programs starting up, the high-value defense R&D contracting marketplace has shrunk significantly across the major DoD components.

Has the budget drawdown led to increasing concentration within the defense industrial base?

To study trends in the concentration of the defense industrial base across the current budget drawdown, CSIS looked at the top 20 prime vendors (measured by prime contract obligations) in 2009 and 2015), and calculated the shares of contract obligations going to the top 5 and top 20 vendors. For DoD overall, the data show no significant shift in the concentration of the defense industrial base overall since 2009.

There has, however, been a significant increase in the concentration of the DoD products market among both the top 5 and top 20 vendors since 2009, with both groups capturing increasing shares of DoD products contracts. The rising concentration in products was offset by the massive decline in the concentration of the DoD R&D industrial base, due to the trough in DoD's development pipeline. Since the largest vendors disproportionately perform the largest R&D projects, it is not surprising that a dearth of large development programs would drastically reduce the share of R&D going to those large prime vendors.

Looking at DoD services vendors, there have been only minor shifts in the composition in the industrial base. This relative stability over the period, particularly over the last two years, may be somewhat surprising given the wave of M&A activity within the government services sector in recent years. The overall trend with the government services market is twofold: first, a trend of diversified vendors divesting their government services business units (particularly in government IT services); and second, of government services-focused vendors merging with or acquiring other vendors to increase market share and access to markets/sectors. These changes, however, have not yet been reflected in the data on the composition of the DoD services industrial base.

What is the future of international cooperative programs given the global populist backlash?

Grounded in a Naval Postgraduate School-sponsored case study of international joint development programs, which established that while defense acquisition is hard international joint development programs are even harder, CSIS held a discussion focused on this question in January 2017. Based on that discussion, there may well be a reduction in top-level support for new multinational programs in the United States, in part because these programs require strong institutional champions to work through their many challenges and such institutional champions are currently not much in evidence. However, for most of the world, the question is not whether to do joint development, but when and how, because most countries simply do not have the capacity and funding to go it alone on highly complex and expensive development efforts. Traditionally, the United States has had the option to go it alone, but even increased U.S. military funding will not eliminate the need to make tradeoffs in the allocation of research and development funding, and the increasingly globalized technology base will serve as a growing incentive to look abroad for partners in developing cutting edge technology. Hence it is likely that the United States also will again see the need to engage in international joint development going forward.

For these future programs, wise design is paramount. Successful international joint development programs should satisfy multiple categories of objectives, be they security, political, or economic, for each participant. There are also specific approaches to mitigate the challenges of multinational projects. Component compartmentalization, splitting the project into discrete severable design elements, can help distribute workshare while mitigating the resulting complexity. Best value subcontracting can help keep costs down. Under the system used by the F-35, partner nations aren't guaranteed workshare, they have to offer a good price, which may mean their home country makes a bigger indirect contribution via subsidies. Finally, joint international development may be a key to cooperation with countries such as India that are interested in developing their own domestic defense industry. If a closer relationship and expanded defense trade remains a policy goal, past history suggests that mutually beneficial arrangements are available, but all involved should acknowledge that technology transfer, like building trust, is a slow-moving process built over multiple successes rather than leaps forward.

What Are the Defense Components Buying?

Has the end of the drawdown in overall DoD contract obligations carried over to the Army?

Of the three military services, the Army has been hit hardest by the declines in DoD contract obligations since 2012, falling by 36 percent over the 2012–2015 period, notably more steeply than overall DoD contract obligations. 2014 saw a 14 percent decline within the Army's contracting portfolio, which was also steeper than the decline for overall DoD, but in 2015, Army contract obligations declined by only 6 percent, which was roughly in line with the overall decline in DoD contract obligations. In 2016, overall Army contract obligations were virtually flat.

The overall slowing of the decline indicates that the steep decline in Army contract obligations since 2009, driven by the winding-down of combat operations in Iraq and Afghanistan and the Army's recent inability to start and sustain major development programs, may be close to reaching its bottom. The overall stability in Army contract obligations in 2016 lends further evidence to support this hypothesis, but with a dearth of new major development or procurement programs in the pipeline, the Army is unlikely to see significant increases in contract obligations for the foreseeable future.

How significantly does the Air Force lag in promoting competition for services contracts?

In an October 2015 report on defense competition, CSIS noted that the Air Force has seen a significant decline in its level of effective competition for services contract obligations, in a period where competition rates for services in other parts of DoD are either stable or rising. For overall Air Force services, the rate of effective competition (54 percent) was already notably lower than the rate for non-Air Force services (70 percent). That gap has only widened in recent years: while the rate of effective competition for non-Air Force services has risen slightly, to 72 percent in 2015, the rate for Air Force services has fallen to 39 percent in 2015.

Even when looking at the specific types of services (such as Maintenance/Repair of Aircraft or Engineering & Technical Services) that the Air Force contracts for, the Air Force consistently sees lower (and often declining) rates of effective competition for the same types of services that does the rest of DoD. This trend continues to surprise the study team, because the Air Force has been seen as taking a leading role in improving tradecraft in the acquisition of services. CSIS urges policy makers to gain a better understanding of why the Air Force is experiencing this consistent, massive, broad-based decline in competition within its services contracting portfolio.

Appendix A: Methodology

For nearly a decade, the Defense-Industrial Initiatives Group (DIIG) has issued a series of analytical reports on federal contract spending for national security across the government.[110] These reports are built on FPDS data, presently downloaded in bulk from USAspending.gov._DIIG now maintains its own database of federal spending, including years 1990–2014, that is a combination of data download from FPDS and legacy DD350 data. For this report, however, the study team primarily relied on FY2000–2015, along with an initial look at the FY2016 data in Chapter 2. Data before FY2000 require mixing sources and incurs limitations discussed in section A.1.

Inherent Restrictions of FPDS

Since the analysis presented in this report relies almost exclusively on FPDS data, it incurs four notable restrictions.

First, contracts awarded as part of overseas contingency operations are not separately classified in FPDS. As a result, we do not distinguish between contracts funded by base budgets and those funded by supplemental appropriations.

Second, FPDS includes only prime contracts, and the separate subcontract database (Federal Subaward Reporting System, FSRS) has historically been radically incomplete; only in the last few years have the subcontract data started to approach required levels of quality and comprehensiveness.[111] Therefore, only prime contract data are included in this report.

Third, reporting regulations require that only unclassified contracts be included in FPDS. We interpret this to mean that few, if any, classified contracts are in the database. For DoD, this omits a substantial amount of total contract spending, perhaps as much as 10 percent. Such omissions are probably most noticeable in R&D contracts.

Finally, classifications of contracts differ between FPDS and individual vendors. For example, some contracts that a vendor may consider as services are labeled as products in FPDS and vice versa. This may cause some discrepancies between vendors' reports and those of the federal government.

Constant Dollars and Fiscal Years

All dollar amounts in this data analysis section are reported as constant FY 2014 dollars unless specifically noted otherwise. Dollar amounts for all years are deflated by the implicit GDP deflator calculated by the U.S. Bureau of Economic Analysis, with FY2014 as the base year,

---

[110] This appendix draws from numerous past Defense Contracting and Federal Services Contracting Reports. See http://csis.org/program/methodology for the latest version of this methodology. When the methods are drawn from new research within this past year, the specific source is noted in the footnotes.

[111] For more on the current quality and comprehensiveness of FSRS, see Nancy Y. Moore, Clifford Grammich, and Judith Mele, "Findings from Existing Data on the Department of Defense Industrial Base," RAND Corporation, 2014.

allowing the CSIS team to more accurately compare and analyze changes in spending across time. Similarly, all compound annual growth values and percentage growth comparisons are based on constant dollars and thus adjusted for inflation.

Due to the native format of FPDS and the ease of comparison with government databases, all references to years conform to the federal fiscal year. FY2014, the most recent complete year in the database, spans from October 1, 2013, to September 30, 2014.

Included Agencies

This report tracks all contracting activity managed by DoD components with exceptions noted here. The civilian portion of U.S. Army Corps of Engineers contracting is also incorporated. However, contracts funded by DoD but managed by other agencies, such as the General Services Administration, are not included except in budget-related charts where DoD *funded* contracts are explicitly referenced. Finally, in FY2013, the Defense Commissary Agency (DeCA) stopped reporting most of its contract obligations (approximately $5 billion) into FPDS. Because this creates a significant data discrepancy that distorts trend analysis, CSIS has excluded DeCA from the dataset throughout the study period.

Data Reliability Notes and Download Dates

Any analysis based on FPDS information is naturally limited by the quality of the underlying data. Several Government Accountability Office (GAO) studies have highlighted the problems of FPDS (for example, William T. Woods' 2003 report "Reliability of Federal Procurement Data," and Katherine V. Schinasi's 2005 report "Improvements Needed for the Federal Procurement Data System—Next Generation").

In addition, FPDS data from past years are continuously updated over time. While FY2007 was long closed, over $100 billion worth of entries for that year were modified in 2010. This explains any discrepancies between the data presented in this report and those in previous editions. The study team changes over prior-year data when a significant change in topline spending is observed in the updates. Tracking these changes does reduce ease of comparison to past years, but the revisions also enable the report to use the best available data and monitor for abuse of updates.

Despite its flaws, the FPDS is the only comprehensive data source of government contracting activity, and it is more than adequate for any analysis focused on trends and order-of-magnitude comparisons. To be transparent about weaknesses in the data, this report consistently describes data that could not be classified due to missing entries or contradictory information as "unlabeled" rather than including it in an "other" category.

The 2015 data used in this report were downloaded in February 2016. The 2016 data used in this report was downloaded in January 2017; a full re-download of all back-year data was performed simultaneously.

A.1 Detailed Methods

The prior sections apply to all DoD contracting data or the data for years 1990 to 1999. The sections below are specific to only selected graphs or tables that posed additional technical challenges.

A.1.1 Comparison between Contract Obligations and Total Obligations

Data for total DoD obligations were obtained from the Financial Summary Tables available for each fiscal year on the website of the under secretary of defense (comptroller), specifically the "Obligations and Unobligated Balances by Appropriations Account" table.

There is, however, a complication to using these data: the "Total Obligations" column double counts reimbursable activity (such as obligations through a Working Capital Fund, WCF), because it captures both the money obligated by the WCF and the money obligated by customers into the WCF. This is no small issue, because "Reimbursable Obligations" have totaled $150–$200 billion in most years during the period observed. To account for this issue, the study team subtracted "Reimbursable Orders" in each fiscal year from "Total Obligations," to produce a new total that CSIS calls "Total Net DoD Obligations." This total allows CSIS to capture the money obligated out of WCFs (which includes significant contracting activity), while eliminating the double-counting from "Reimbursable Orders," which represents the money paid into the WCFs.[112]

Obligations for the Army Corps of Engineers (ACE) and foreign military sales through the Defense Security Cooperation Agency (DSCA) are not included in the totals referenced above, but significant contracting activity is performed under those two agencies. To allow for a true like-to-like comparison of contract obligations to total obligations, total net obligations for the ACE and DSCA are added to and included in "Total Net DoD Obligations." While ACE accounted for roughly $8 billion to $12 billion in net obligations during the period, DSCA net obligations varied widely from year to year, accounting for over $20 billion in one year, and as little as -$5 billion in another (due to reimbursements outweighing obligations).

A.1.2 Competition[113]

The study team followed DoD methodology and calculated competition by using two fields: extent of competition, which is preferred for contract awards; and fair opportunity, which is preferred for task and delivery orders under most indefinite delivery vehicles (IDVs). In the vast majority of cases, competitive status is classified for the entire contract duration. Thus, if a contract had a duration of three years and was competed in the first year, it qualifies as competed for the entire duration. This also extends to single-award indefinite delivery contracts, which are classified based on whether the original vehicle was competed rather than consistently treated as only receiving an offer from the single awardee. However, for

---

[112] Note that the totals for "Reimbursable Orders" and "Reimbursable Obligations" are not equal in a given fiscal year, due to time discontinuities between obligations by the WCF and orders by the WCF's customers.
[113] This section is adapted from Sanders, *Avoiding Terminations, Single-Offer Competition, and Costly Changes with Fixed-Price Contracts.*

some other vehicles, such as multiple-award IDVs, the number of offers is instead tracked separately for each task order.

To better evaluate the rate of "effective competition," the study team categorizes competitively awarded contracts by the number of offers received.[114] CSIS focuses on the number of offers for competed contracts because it reveals information about the request for proposals. A solicitation that only has a single respondent indicates some combination of three factors: thinness in the underlying market; a failure to notify or give adequate response time to potential competitors; or a contract that is unappealing to vendors.

The focus on the number of offers also has a basis in the regulation known as the Single Offer rule (DFARS 215.371), which addresses competitive acquisitions in which only one offer is received. This rule was rewritten in 2012 to add a policy section that shifts emphasis away from an analysis of whether the circumstances described at FAR 15.403-1 (c)(1)(ii) (determining adequate price competition) are present, to whether statutory requirements for obtaining certified cost or pricing data are met and if the price is fair and reasonable. The revised rule also emphasizes the need to extend the period of solicitation when only one offer is received, to see whether a longer response period can elicit additional bids. Essentially, the new standard suggests that if you cannot get two bidders, you must evaluate whether proceeding with one bid can be done while protecting the interests of the government.

A.1.3 Contract Initial Duration and Size[115]

When contract initial duration and size become factors, the dataset used is limited to contracts reported in FPDS that were initially signed no earlier than FY2007 and completed by FY2013. Determining when contracts are completed is the most challenging portion of compiling the dataset. Contracts closed out or terminated by the end of FY2013 are included even if their current completion dates run into the next fiscal year. However, many contracts in FPDS and in the sample are never marked as closed out or terminated in the Reason for Modification field. In these cases, completion status is based on the current completion date of the most recent transaction in FPDS. This method could accidentally include contracts that have not reached their ultimate conclusion dates and are merely dormant. However, the FY2013 sample end date means that any such contracts would have to be inactive for an entire fiscal year, which is unlikely.

FPDS raw data are available in bulk from USAspending.gov starting in FY2000. However, data quality steadily improves over that decade and a half, particularly in the commonly referenced fields of interest to this study. In most cases, unlabeled rates topped out at 5 to 10 percent. The critical exceptions are the Base and All Options and Base and Exercised Options fields, which report contract ceilings. Prior to FY2007, these fields are blank for the majority of contracts. When that field is not available, calculating the extent of ceiling breaches is

---

[114] CSIS defines effective competition as a competitively sourced contract awarded after receiving two or more offers.

[115] This section is adapted from Sanders, *Avoiding Terminations, Single-Offer Competition, and Costly Changes with Fixed-Price Contracts*.

impossible. In addition, this study classifies contract size by original ceiling and not total obligations because the latter figure is dependent on contract performance.

Because a key dependent and independent variable are not available prior to FY2007, the study team chose to set FY2007 as the start date rather than risk sample bias by including only those earlier contracts that were properly labeled. This restriction poses a significant limitation in that no contracts of more than seven years in duration can be included and five-year contracts are only in the study period if they started by October 1, 2007, or were closed out early.

A.1.4 Terminations[116]

Contract termination is determined through the Reason for Modification field in FPDS. A contract is considered terminated if it has at least one modification with the following values:

- "Terminate for Default (complete or partial)"

- "Terminate for Cause"

- "Terminate for Convenience (complete or partial)"

- "Legal Contract Cancellation"

These four categories and the "Close Out" category are used to mark a contract as closed. As discussed above, many contracts well past their current completion date never have a transaction marking them as closed; however, a termination is an active measure that mandates reporting, unlike the natural end of a contract, which can go unremarked.

The four different values of contract termination provide useful granularity, but even a termination for convenience indicates that something has likely gone awry. Thus, given the already low number of terminations, the study team treats a contract as either terminated or not, rather than subdividing by type.

A.1.5 Change Orders and Ceiling Breaches[117]

Similar to contract terminations, change orders are reported in the Reason for Modification field. There are two values that this study counts as change orders: "Change Order" and "Definitize Change Order." For the remainder of this report, contracts with at least one change order are called Changed Contracts.

There are also multiple modifications captured in FPDS that this current study will not investigate as change orders. These include:

- Additional work (new agreement, FAR part 6 applies)

---

[116] Ibid.
[117] Ibid.

- Supplemental agreement for work within scope

- Exercise an option

- Definitize letter contract

The Number of Change Orders refers to the number of FPDS transactions for a given contract that lists one of the two change order categories as their Reason for Modification. The vast majority of contracts do not receive change orders, but changed contracts are still far more common than terminations.

The study team calls when the total potential cost of a contract increases due to a change order ceiling breach. In federal acquisition, the government usually sets a "cost ceiling" of contracts that limits the total amount of funds it may obligate on a single contract. This maximum cost ceiling can serve as a target for vendors looking to maximize their revenue under a contract. However, cost ceilings can be raised, meaning that they do not represent true maximums. When work under a contract is set to exceed the contract ceiling for any reason, the government is forced to breach these cost ceilings. "Ceiling Breaches" represent output indicators, because they indicate that either the real cost of a contract or its true scope of work was not fully understood at the time of contract award.

This study uses changes in the Base and All Options Value Amount as a way of tracking the potential cost of change orders. The Base and All Options Value Amount refers to the ceiling of contract costs if all available options were exercised. The alternative ceiling measure, Base and Exercised Value Amount, is not used because contracts are often specified such that the bulk of the eventually executed contract, in dollar terms, is treated as options. In these cases, the all-inclusive value provides a better baseline for tracking growth.

The Obligated Amount refers to the actual amount paid to vendors. This study team does not use this value for the analysis because spending for change orders is not necessarily front-loaded. For example, a change to a contract in May 2010 could easily result in payments from May 2010 through August 2013.

A.1.6 Vendor Categorization

Small, Medium, and Large Vendors

To analyze the breakdown of competitors in the market into small, medium, and large vendors, the CSIS team assigned each vendor in the database to one of these size categories. Any organization designated as small by the FPDS database—according to the criteria established by the federal government—was categorized as such unless the vendor was a known subsidiary of a larger entity. Due to varying standards across sectors, an organization may meet the criteria for being a small business in certain contract actions and not in others. The study team did not override these inconsistent entries when calculating the distribution of value by vendor size.

Vendors with annual revenue of more than $3 billion, including from nonfederal sources, are classified as large. This classification is based on the vendor's most recent revenue figure at time of classification. For vendors that have gone out of business or been acquired, this date may be well before 2014. A joint venture between two or more organizations is treated as a single separate entity, and organizations with a large parent are also defined as large. Due to their system integrator role and consistent market share, the study team placed the five largest defense contractors (Lockheed Martin, Boeing, Raytheon, Northrop Grumman, and General Dynamics) into a separate category called "Big 5 defense vendors." Any vendor assigned a unique identifier by FPDS but is neither small nor large is classified as "medium."

To identify large vendors, the study team investigated any vendor with total obligations of $500 million in a single year or $2 billion over the study period. Determining revenues is the most labor-intensive part of the process and involves the use of vendor websites, news articles, various databases, and public financial documents. When taken together, all of this work explains the increase in the market share of large vendors versus some older editions of this report. While large vendors are, on rare occasions, reassigned into the middle tier, the vast majority of investigations either maintain the status quo or identify small or medium vendors that should be classified as large.

Handling of Subsidiaries and Mergers and Acquisitions

To better analyze the defense industrial base, the study team made significant efforts to consolidate data related to subsidiaries and newly acquired vendors with their parent vendors. This results in, among other things, a parent vendor appearing once on CSIS's top 20 lists rather than being divided between multiple entries. The assignment of subsidiaries and mergers to parent vendor is done on an annual basis, and a merger must be completed by the end of March in order to be consolidated for the fiscal year in question. This enabled the study team to more accurately analyze the defense industrial base, the number of players in it, and the players' level of activity.

Over the past seven years, the study team has applied a systematic approach to vendor rollups. FPDS uses hundreds of thousands of nine-digit DUNS (Data Universal Numbering System) codes from Dun and Bradstreet to identify service providers. A salutary benefit of this standardization is that FPDS now provides parent vendor codes. These parent codes track the current ownership of vendors but are not backward looking. Thus, a merger that happened in 2010 would not affect parent assignments in 2000. This prevents the study team from adopting these assignments in their entirety. The study team investigates vendors that receive $250 million of total contract revenue or more than $1 billion in obligations between 2000 and 2014, no matter how much they receive in any individual year. We have reinforced these manual DUNS number assignments with automated assignments based on vendor names. Qualifying for an automated assignment by name requires three criteria: 1) a standardized vendor name that matches with the name of a parent vendor, 2) that the name has been matched to the parent vendor by the CSIS or the Parent DUNS number field, and 3) there are no alternative CSIS assignments with that vendor name. This process is not immune to error, but it reduces the risk that a DUNS code is considered large in one year but overlooked in another. As an error-checking mechanism, the study team investigated contradictions by comparing our assignments to those made by Parent DUNS numbers for every DUNS number with $500 million in annual obligations or $2 billion in total obligations.

About the Project Directors and Authors

Andrew Hunter is a senior fellow in the International Security Program and director of the Defense-Industrial Initiatives Group at CSIS. From 2011 to 2014, he served as a senior executive in the Department of Defense, serving first as chief of staff to undersecretaries of defense (AT&L) Ashton B. Carter and Frank Kendall, before directing the Joint Rapid Acquisition Cell. From 2005 to 2011, Mr. Hunter served as a professional staff member of the House Armed Services Committee. Mr. Hunter holds an M.A. degree in applied economics from the Johns Hopkins University and a B.A. in social studies from Harvard University.

Rhys McCormick is a research associate with the Defense-Industrial Initiatives Group (DIIG) at CSIS. His work focuses on unmanned systems, global defense industrial base issues, and U.S. federal and defense contracting trends. Prior to working at DIIG, he interned at the Abshire-Inamori Leadership Academy at CSIS and the Peacekeeping and Stability Operations Institute at the U.S. Army War College. He holds a B.S. in security and risk analysis from the Pennsylvania State University and an M.A. in security studies from Georgetown University.

Jesse Ellman is a research associate with the Defense-Industrial Initiatives Group (DIIG) at CSIS. He specializes in U.S. defense acquisition policy, with a particular focus on Department of Defense, Department of Homeland Security, and government-wide services contracting trends; sourcing policy and cost estimation methodologies; and recent U.S. Army modernization efforts. Mr. Ellman holds a B.A. in political science from Stony Brook University and an M.A. with honors in security studies, with a concentration in military operations, from Georgetown University.

Samantha Cohen *was* a research assistant with the Defense-Industrial Initiatives Group at CSIS. Her work focused on managing and analyzing data to identify relationships among policies, defense spending, and outcomes. Her recent research focused on public opinion and defense spending in European countries. In 2015, she graduated from American University with a B.S. (honors) in economics *and is currently pursuing an M.S. in economics at the University of Leuven*. Additionally, she successfully completed the intensive French Language School program at Middlebury College and studied the economic and defense policies of the European Union and NATO with AU at the Université catholique de Louvain in Brussels, Belgium, in 2014.

Kaitlyn Johnson is a program manager and research associate for the Defense-Industrial Initiatives Group (DIIG) and a research associate for the Aerospace Security Project at CSIS. Her work focuses on supporting DIIG research staff, as well as specializing in research on defense acquisition and space policy. Previously she has written on ultra-low cost access to space, defense acquisition trends, and federal research and development contract trends. Ms. Johnson holds an M.A. from American University in U.S. foreign policy and national security studies with a concentration in defense and space security, and a B.S. from the Georgia Institute of Technology in international affairs.

Gregory Sanders is a fellow with the International Security Program and deputy director for research of the Defense-Industrial Initiatives Group at CSIS, where he manages a team that analyzes U.S. defense acquisition issues. Utilizing data visualization and other methods, his research focuses on extrapolating trends within government contracting. This requires innovative management of millions of unique data from a variety of databases—most notably the Federal Procurement Database System, and extensive cross-referencing of multiple budget data sources. Mr. Sanders holds an M.A. in international studies from the University of Denver and a B.A. in government and politics, as well as a B.S. in computer science, from the University of Maryland.